FATE, FREEDOM, AND INFINITY

ESCAPING THE DEVIL'S GARDEN

CASSIUS JACKSON KESYER

FATE, FREEDOM, AND INFINITY

ESCAPING THE DEVIL'S GARDEN

CASSIUS JACKSON KESYER

Edited by Tristan Abbey and Kevin M. Kambo

Science Venerable Press
Tampa, Florida
MMXXIII

Cover art: White paper origami flower and origami detail/iStock by Getty Images

ISBN 978-1-959976-04-2

Published by Science Venerable Press, an imprint of Comarus Analytics LLC
Tampa, Florida
Sciencevenerable.com

Interior designed by Emily Weigel
Cover designed by Kathleen Lynch

"Wonder is the great question-asker.
Answers are propositions."

CASSIUS JACKSON KEYSER

PREFACE

Cassius Jackson Keyser was a tedious writer. More often than not, the pages blur together and the word count ascends without limit. The jargon itself is an agony. This project is concerned entirely with those other occasions—when his poetic connections between mathematics and philosophy set the text ablaze.

This work is an amalgam of three free-standing published works authored by Cassius Jackson Keyser during the 1915-1939 period. Our core is a collection of biographical sketches originally printed alongside a series of portraits. All the portraits have been replaced with updated artwork. The original text of the sketches has been preserved nearly in full. This edition provides new annotations (Keyser included none) to trace source material, cleans up quotations that are frequently inaccurate or rely on a mix of jumbled translations, modernizes spelling and formatting (primarily punctuation), and removes miniature in-line bibliographies that added value a century ago but are a distraction today. The introduction and conclusion are extracted from separate works chosen purely because they seem to fit together.

Edits are intended to enhance the reading experience. In all cases, [bracketed] notes indicate our own editorial comment. The initials C.J.K. have their obvious meaning. We have also provided a set of discussion questions and a detailed References section at the end.

In Francis Bacon's *New Atlantis*, the technologically advanced city Bensalem ("Son of Peace" or "Son of Submission") is ruled by a science institute called The College of the Six Days' Works, whose name bespeaks godlike power to order and govern the cosmos. In the College, intellectual pioneers are celebrated with religious devotion: "We have two very long and fair galleries, in one of these we place patterns and samples of all manner of the more rare and excellent inventions, in the other we place the statues of all principal inventors... For upon every invention of value, we erect a statue to the inventor and give him a liberal and honorable reward." Scientific inventions are treated as intellectual relics, and the great scientists who discovered them are venerated as intellectual saints. In the writings of Keyser, similarly, we find a gallery of verbal portraits, literary statuary to the communion of mathematical sages.

For Keyser, mathematics is not simply a numerical or geometric activity; it is a spiritual exercise. In these meditations on intellectual figures through history, one finds firm interest—indeed, belief—in a chivalric dimension to science and mathematics; there is romance in the pursuit of truth, the quest to understand the cosmos. As such, the intellectual life (and its history) is not simply the story of big brains working out problems algorithmically; it is a drama featuring the full palette of human motivations, conflicts, and (short-lived) resolutions.

One gazes upon Keyser's portraits, therefore, and is drawn into fascinating questions. To what extent, for example, can Keyser's mathematical saints be said to belong to a single, coherent tradition? Some of the figures, like Keyser, are devoted to the ancients, while others are keen to present themselves as independent thinkers. Some authored systems, others did not. They disagree on the ultimate authority of mathematics among the sciences. They have different responses to the limits of human knowing, as well as to the twin threats to knowledge—rigid dogmatism and radical skepticism. Through Keyser's catalogue of mathematical legends, then, one is invited to think richly about our intellectual inheritance.

We make no pretensions of having produced either a biography (critical or hagiographic) or scholarly assessment of Keyser's contributions to mathematics and philosophy. Cassius Jackson Keyser simply had interesting things to say about interesting things, and many of these things still interest us today.

TRISTAN ABBEY
Tampa, Florida

KEVIN M. KAMBO
Irving, Texas

TABLE OF CONTENTS

CASSIUS JACKSON KEYSER

INTRODUCTION

Mathematics contemplates Being under the aspect of Eternity; for it is perfectly clear that doctrinal functions, though their discovery by man is a temporal event, are themselves timeless—"older than the Sun or the Sky" and destined to survive all things that are under the law of changes and the doom of death.

The view I have been presenting, in which pure mathematics appears as a vast array of doctrinal functions, gives the science, from one point of view, a pretty severe aspect. For a doctrinal function is not only timeless, as said, and indestructible, but—and the fact merits our most pensive meditation—when, once the principles, or postulates, are chosen, the die is cast—all else follows with a necessity, a compulsion, an inevitability that are absolute—we are at once subject to a destiny of consequences which no man nor any hero nor Zeus nor Yahweh nor any god can halt, annul or circumvent.

Mathematics is, in a world, the study of Fate.

Let me hasten to say that the Fate is not physical, it is spiritual—the unbreakable binding thread of destiny runs through the universum of rigorous Thought: the fate of logical Fate. Is it a tyrant? And the intellect, then, a slave? A tyrant has whims but Logic is awful. Where, then, is the intellect's freedom? What do you love? Poetry? Painting? Architecture? Statuary? Music? The muses are their fates. If you love them, you are free. Logic is the muse of thought. When I violate it, I am erratic; if I hate it, I am licentious or dissolute; if I love it, I am free—the highest blessing the austerest muse can give.

I am using the term Thought to signify that sort of discourse which deliberately owns allegiance, even though it often fails in loyalty, to the authority of Logic. The subject is, you see, immense, penetrating all the sciences and all the philosophies, natural, or social, or speculative—all fields, in short, where men have sought by means of reasoned discourse to gain or to give wisdom and light for the guidance of humankind.

"God Himself,
the father and
fashioner of all that is,
older than the Sun
or the Sky, greater
than time and eternity
and all the flow of
being, is unnameable
by any lawgiver,
unutterable by
any voice, not to be
seen by any eye."

MAXIMUS OF TYRE,
2ND CENTURY A.D.

We humans desire indeed to be regarded devoted lovers of truth and we flatter ourselves that we are such in fact; sometimes we are, but, in general, we are not; in general, we prefer something else; we often boast that we are not theoreticians, and the boast has its basis in fact; we are not theoreticians, we are practicians, though we dislike the word and call ourselves practical instead. Being primarily and predominantly practical in our interests, when we are building doctrines, though we always pretend to be thus endeavoring to set forth truth, we are, with rare exceptions, animated by a very different motive; we are not trying to formulate something that, by painstaking research, we have found to be true; we are, instead, though we do not confess and may not even know our real motive, trying to make an instrument that will "work," that will be—an effective means to some practical end; we are not, however much we pretend to be—endeavoring to enlighten our fellow men—we are endeavoring to influence them: our aim is not the advancement of wisdom; it is, in current slang, to put something over or across.

Well, you may say, what is to be done about it? What is the remedy? The remedy is—Criticism—the Gadfly: patient, unsparing logical criticism of one's own work in doctrine building; and, in all subjects, keen, merciless, stinging, gadfly criticism of any and all half- baked, logically amorphous, flabby doctrines pretending to be important embodiments of truth or vessels of wisdom. Men must be driven by art, the art of criticism, to levels of excellence higher than those to which they are drawn by unenlightened nature.

Consider the propositional function: if x functions a kind y of activity, then x has the property z. Let x denote I, let y denote thinking, let z denote being, drop the hypothetical form, and you have Descartes: "Cogito, ergo sum; Je pense, donc je suis; I think, therefore I am." Enough of hints. The suggested type of analysis is evidently applicable on every hand—to the Sermon on the Mount [Matthew 5-7], to the Republic of Plato, to Darwin's Origin of Species [1859], to the League of Nations Covenant [1919], to Marxian Socialism, to the Soviet Constitution of Russia [1918], to the Constitution of the German Republic [1919], to the Einstein Doctrine of Relativity, to the Bryanistic Ethics of Prohibition[1] to all manner of doctrinistic contentions of wise men, knaves, fanatics and fools.

The type of criticism I am here advocating and urging as supremely important shapes itself, as you see, very simply. Confronted by a doctrine in any department of thought, Criticism demands answers to these questions: What is assumed—what are the postulates? What are the undefined, or variable, terms? What are the theorems or proved propositions and what the defined, or constant, terms? How have the theorems been deduced, and the defined terms defined? What meanings have been assigned to the variable terms, and how? Upon these questions, criticism, if it is to be criticism of Thought, is bound to insist—there is no alternative. Such criticism is a civilizing agency—the guardian of the principles of freedom. Without it, the world becomes a wilderness of error and lust—the garden of the Devil.

Easy to ask, the questions are, in general, not easy to answer, and the difficulty of answering rightly is usually greatest—just where it is most important to compel an answer in the case, that is, of amorphous, emotion-charged

1 William Jennings Bryan (1860-1925) supported the constitutional prohibition of alcohol, ratified in 1919 and repealed in 1933.

"This is the way that weird geometry proceeds: it sets before beginners certain strange assumptions, and insists on their granting the existence of inconceivable things, such as points having no parts, lines without breadth, and so on, builds on these rotten foundations a superstructure equally rotten, and pretends to go on to a demonstration which is true, though it starts from premises which are false."

LUCIAN OF SAMOSATA, 2ND CENTURY A.D.

"dynamic" doctrines that pretend to aim at enlightenment but really aim at victory and win it by appealing, not to love of truth, but to lust for power or gain. If the author be unable to answer, criticism must drive him back to the silence of the cloister for further study. If he contends, as sometimes he will contend, that he has defined all his terms and proved all his propositions, then either he is a performer of logical miracles or he is an ass; and, as you know, logical miracles are impossible.

It is necessary to distinguish: to test for truth is one thing; to utter is another. Of these two things, the former is the duty to which men must be driven by criticism if they be not drawn to it by love of truth. Once the test is made and the doctrine found not wanting, then, and not before, it may be legitimately urged home with full ardor by all the arts of utterance even though the truth be thus made to burst upon us like the thunder of Wagnerian music,[2] making the mountains tremble, the seas vibrate, and seeming to shake the very rafters of the sky.

2 Wagner's most famous work, *The Ring of the Nibelung*, premiered at the New York Metropolitan Opera in 1889. C.J.K. easily could have attended.

About a generation and a half ago a distinguished American mathematician, on discovering that one of his students though primarily interested in mathematics had also a lively interest in philosophy, said to him: "Mathematics and philosophy are incompatible; the effect of mathematics on philosophy is like that of sulfur on the itch."[3] That foolish speech could not have been made by one who knew enough of the history of thought to know that some of the world's most eminent philosophers were also eminent mathematicians.

That all of the thinkers presented in this portfolio achieved great renown as philosophers will not be disputed. It is not so generally known that all but one of them were well qualified by their scholarship, insightfulness, sympathy, and understanding to be ranked also as mathematicians, some of them as creative mathematicians of the highest order. "All but one," I have said. That one is Epicurus. Epicurus was not only no mathematician but, like Lucian of a much later time, he held mathematics in contempt. Why, then, is he included here? Because in constructing his philosophy Epicurus made bold and brilliant, though sometimes very erratic, use of a stupendous concept which was destined to become, after the lapse of more than two thousand years, a central pillar in the majestic edifice of modern mathematics. I mean the concept of Infinity.

The following brief biographical accounts have not been set down hastily, but only after much reading and pondering of what the heroes themselves wrote and of what others have said of them. One exception to this statement must be noted. For Pythagoras reduced no part of his thought to writing but communicated his ideas to his disciples solely by means of the spoken word, and so in reporting upon his life, personality, discoveries, and teachings one is obliged to depend upon the findings of those scholars who have critically examined the great Pythagorean tradition.

3 Sulfur remains an ingredient of many anti-itching ointments.

PYTHAGORAS

1

PYTHAGORAS

About 582-500 B.C.

It is impossible to write what could properly be called *The Life of Pythagoras*. For not enough is known or ascertainable about the facts of his life; no one can say, with more than a rough approximation to completeness or accuracy, which of the many discoveries traditionally credited to Pythagoras were really his; nor is it possible to know exactly either the content or the form of his teachings. The reasons, which are simple, are these:

(1) Pythagoras lived and died long before the Greeks were sufficiently interested or had learned so to record the constituent facts of a great man's life that, at his death or in after time, they could be available as material or as a basis for authentic biography. The writing of history had not yet begun. The birth of Pythagoras antedated that of the Father of History [Herodotus] by at least a hundred years, and in antiquity, when the wastebasket of memory was the sole means of keeping the record of a human life, a century was a long stretch of time. It is, then, not a matter for wonder that the three ancient so-called *Lives of Pythagoras*, by Porphyry, Iamblichus, and Diogenes Laërtius, composed seven hundred to eight hundred years after their subject's death, are little better than "a mass of incredible fiction." Far more trustworthy are Plato and Aristotle. They say a good deal about the Pythagoreans, as the disciples of Pythagoras were called, but very little about Pythagoras himself. The *Phaedo* and the *Timaeus* of Plato are far more than deeply tinged with Pythagorean ideas, and Aristotle was sufficiently interested in the Pythagoreans to write a now lost treatise about them. Pythagoras himself is mentioned by Plato only once, as a man beloved by his disciples because he had taught them "a [certain] way of life." By Aristotle he is

"In consequence of having divulged and described the method of forming a sphere from twelve pentagons, Hippasus perished in the sea, as an impious person, but obtained the renown of having made the discovery."

IAMBLICHUS, ***LIFE OF PYTHAGORAS***

mentioned twice.[4] But none of the three mentions adds aught to our knowledge of biographic fact.

(2) It appears that Pythagoras and, after his death, his memory were so revered by his early disciples that none of them could, without great impiety, claim for his own the glory of any discovery but all must be referred back to the master—to Him, as he was designated without calling his name. It is said that Hippasus, the Pythagorean, was drowned at sea because he impiously laid claim to having originated the spherical inscription of a dodecahedron [12-sided solid]. Moreover, like Thales before him, neither Pythagoras nor his early disciples reduced their ideas to writing but communicated them solely by word of mouth. Consequently, though the ancient writings about Pythagoreanism contain many statements ascribed to Pythagoras, none of them can be confidently quoted as his, either in content or in form.

Evidently, we have to depend exclusively on hearsay and the diverse reports

---

4 Ancient authors invoke Aristotle as a source for fragmentary information about Pythagoras—that he refused to eat beans (Diogenes Laërtius), would say certain things (Porphyry), and had a golden thigh, bit a snake to death, and could prompt a river to speak (Apollonius).

thereof in the great Pythagorean tradition. Yet so vast is that tradition, so widespread, so persistent through the centuries, and so rich in significant components, that, despite its many fictional elements and its obvious lack of perfect self-consistency, one can hardly contemplate it candidly and open-mindedly without descrying by its somewhat misty and flaring light the presence back yonder in the sixth century B.C. of a dim, shadowy, august, polydynamic figure whose relation to the mathematics, the philosophy, and the science of the Classical Culture of ancient Greece deeply resembles the well-known relation of René Descartes to the mathematics, the philosophy, and the science of the Western Culture of the modern world.

Respecting more specific information regarding the life and achievements of Pythagoras, the minute and critical studies of the Pythagorean tradition enable one to report, not wholly without conjecture but with sufficient confidence, as follows.

According to the weight of the evidence, Pythagoras was born about 582 B.C. at Samos and there passed the years of his youth and early manhood. Of his mother nothing is known; of his father, only the name, Mnesarchus. Respecting his education we know only that he was a pupil of Pherecydes, Hermodamas, and, possibly, Thales. Tradition has it that he left Samos to escape the tyranny of the king, Polycrates. For the many years before he established himself about 529 B.C. at Crotona, a Greek city in southern Italy, we have no definite knowledge of his whereabouts or activities. Tradition filled the gap by representing Pythagoras as an extensive traveler, visiting Egypt, Phoenicia, Chaldea, Palestine, Arabia, Gaul, Persia, India. The tales, though doubted, have not been disproved. There are three reasons for thinking they may be true: (1) There was ample time for such journeys; (2) traveling was then perhaps the best way for acquiring knowledge; (3) Pythagorean teachings are often deeply stained with oriental mysticism.

At Crotona, a wealthy commercial city and seaport, Pythagoras established what was then a new kind of society—a secret order or brotherhood—composed of a large number of young men—some say as many as three hundred—chosen from prosperous families. The membership was divided into two classes, the "probationers" and the "mathematicians." According to most of the tradition the fraternity's aim was the attainment and spread of holiness. Such it may have been, for in the sixth century B.C. Greece was deeply stirred by a religious

"As He set in order the Heaven, of that Eternity which abides in unity He made an eternal image, moving according to number, even that which we have named Time."

PLATO,
TIMAEUS

revival. By Oswald Spengler the character of Pythagoras as a prophet has been likened, perhaps insightfully, to that of [the Prophet] Muhammad.[5] But one tradition has it that Pythagoras was a statesman rather than a religious prophet, and another that he and his disciples were primarily interested in science. The traditions need not be thought irreconcilable. For modern specialization did not then exist, and a versatile genius could be at once poet, prophet, theologian, mathematician, scientist, statesman, and reformer of morals. Pure thinking—mathematics, philosophy, science—seems to have been regarded by Pythagoras and his school, as later by Plato, as the best way to "purify the soul" and so release it from the "wheel of birth."

Crotona became the center for branch fraternities in the other cities of Magna Graecia.[6] After twenty years, for a reason not clear, a reaction came and the lodges of the fraternities were violently exterminated, many of the brethren being killed. Pythagoras himself escaped and went to Metapontum where he died near the beginning of the fifth century at the age of about 80. The effect of the

5 Spengler groups together Pythagoras, Muhammad, and Oliver Cromwell.

6 Greater Greece: colonies in southern Italy, such as Syracuse, home of Archimedes.

persecution in Magna Graecia was to propagate the Pythagorean philosophy in Greece proper. As a school of thought Pythagoreanism continued to the middle of the fourth century B.C.

Among the specific achievements to be accredited to Pythagoras, according to the weight of authority, are the following:

(1) The elevation of European mathematics to its proper level above the merely practical and experimental by insisting upon the method of axioms, postulates, definitions, and proofs.

(2) The discovery of irrational numbers by demonstrating the incommensurability of the diagonal and the side of a square.

(3) The discovery and proof of the angle-sum of a triangle (according to surviving fragments of a lost History of Geometry by Eudemus [of Rhodes], c. 335 B.C.) and (on the same authority) the construction of the regular polyhedrons.

(4) The introduction into Greek mathematics and a general proof of the famous theorem about the squares on the hypotenuse and sides of a right triangle.

(5) The proof (though not the discovery) of the theorem that the plane space about a point of it is filled by six regular triangles or four squares or three regular hexagons.

(6) Construction of a polygon equal to another and similar to a third.

(7) Conception of the earth as a globe. Subsequently Pythagorean disciples held the earth to be a planet revolving with the other planets, the moon, the sun, the heaven of the fixed stars, and an invisible counter-earth about a central fire: a doctrine which, says Copernicus, suggested to him his heliocentric hypothesis.[7]

(8) Founding of musical science by discovering the rational divisions of a musical string into intervals yielding the fourth, the fifth, and the octave of a note.

---

7 In *On the Revolutions of the Heavenly Spheres*, Copernicus cites two Pythagoreans—Philolaus and Ecphantus—who believed in "the mobility of the earth."

(9) The coining of the term mathematics and the term philosophy.

(10) Combination of arithmetic and geometry, reminding us of the marriage, two thousand years later, of geometry with algebra.

(11) The attempt to construct a cosmology based on the assumption that the "beginnings" or elements or essences of all things are numbers—a speculation found later in the *Timaeus* of Plato.

(12) Supporting and refining the then old, crude, and widespread *religious beliefs* in the immortality and the transmigration of the soul by philosophically reasoned *doctrines* of lasting influence in the further development of Greek speculative thought.

(13) So fathering the Pythagorean Society that it became the chief scientific school of Greece and that to its work in Geometry we owe the main part of the first two books of Euclid's *Elements*.

To the Pythagoreans tradition ascribes a vast amount of what seems to us numerological nonsense. One sympathizes with Galileo who did not believe this nonsense to be the authentic outgiving of genuine representatives of the school.

The impact of Pythagoras upon the Greek mind of the sixth century B.C. is still felt in philosophy, in religion, in music, in science, and in mathematics.

PLATO

2

PLATO

427-347 B.C.

WITH ONE POSSIBLE EXCEPTION, THE most famous, without any exception the most winged, most charming, and best beloved, of all the philosophers of the Western World. Whether there has been a greater or an equal among the sages of the Orient is a question that may never be answered satisfactorily.

Just as in the case of Euclid, we distinguish between the man and the mathematician, and just as we may know the mathematician well, because he is well revealed in his *Elements* and other extant works, but have to own that respecting the man we know almost nothing, so an analogous situation confronts us in the case of Plato. For there is the philosopher and there was the man. Of Plato's inner life, the life of the philosopher, his copious writings, all of which have come down to us, constitute a sufficiently full and authentic revelation. But regarding the facts of his outward life, the life of Plato the man, our knowledge is exceedingly meager. For, though reports of such facts abound in the Platonic tradition, only a very few of them have withstood the scrutiny of critical scholarship, the rest being now regarded as belonging, not to history, but to the literature of fiction, fable, legend, and myth. The tales are interesting, many of them beautiful, some of them memorable, all of them more or less significant as signs or tokens of something extraordinary in their hero's life, but they afford us no definite knowledge of biographic fact.

We have, then, in speaking of Plato's life, to guard against the danger of mistaking fiction for fact, myth for history, conjecture for knowledge, inferences for data, and the more or less probable for the well ascertained.

It may be regarded as certain that Plato was born in or near Athens, in May, 427 B.C., and attained the age of eighty; that his father, Ariston, and his mother were both of them of distinguished ancestry; and that he had two brothers, Glaucon and Adeimantus, remembered only because he made them participants in two of his

"Plato, when he saw his death approaching, thanked the guiding providence and fortune of his life, first, that he was born a man and a Grecian, not a barbarian or a brute, and next, that he happened to live in Socrates' age."

PLUTARCH,
2ND CENTURY A.D.

greatest dialogues, the *Republic* and the *Parmenides*.

Respecting the time of Plato's birth, two facts are extremely noteworthy: (1) He was born in the wake of the great Age of Pericles—the age of the Greek Enlightenment—perhaps never surpassed—when the intellect of Athens, owing to the presence there of thinkers and teachers from all parts of Greece, had been aflame for forty years with the fire of free enquiry and free expression regarding all things of human interest; (2) he was born three years after the Peloponnesian War began and was twenty-seven when it ended with the humiliation of his native city.

In the absence of definite knowledge it has always been reasonably assumed that Plato's family was at least well-to-do, perhaps wealthy, and that he and his brothers were well schooled in grammar, music, and gymnastics in accord with their station and the educational standards of the time.

At or about the age of twenty, Plato met Socrates, son of a statuary and a midwife. That was a truly momentous event, not only the most momentous event in the course of Plato's education, but among the most momentous in the annals of philosophy. For eight years Plato was of the company of those, mostly young men, who attended Socrates as this snub-nosed, pop-eyed, pot-bellied, thick-lipped, physically ugly but spiritually beautiful man daily went forth in quest of wisdom to one or another of the haunts of men—the agora, a gymnasium, or other place, a public street, perhaps—seeking out such as were reputed wise or were wise in their own conceit, engaging them in conversation and, always insisting that he himself was without knowledge but was eager to learn, questioning them searchingly regarding the special knowledge which they professed or were alleged to have.

When Plato was twenty-eight, his beloved master was put to death. At the age of seventy the great catechiser of the "wise" was brought to trial on the trumped-up charge of "denying the gods recognized by the state, introducing other gods, and corrupting the young," was convicted, sentenced to die, kept for thirty days in prison where he was permitted to receive and converse with his friends as usual. And then, in the name of patriotism, ethics, and religion, "the wisest, the justest, and best of men" was executed—executed by a city that he had done more, perhaps, than any other to make the most intellectually radiant city of all time. The account of all this as given by Plato in the *Apology*, the *Phaedo*,

"Plato is philosophy, and philosophy Plato–at once the glory and the shame of mankind, since neither Saxon nor Roman has availed to add any idea to his categories."

RALPH WALDO EMERSON

and the *Crito* is probably unsurpassed by anything in the realm of dramatic literature. There is a report that Plato used to say: "I thank God that I was born in the age of Socrates." The report is credible.

Of Socrates we have two famous literary portraits, one by Xenophon in his *Memorabilia* and one by Plato in those dialogues in which Socrates is the dominant personality. Both pictures are convincing, for, though they differ a good deal, they differ less than their makers—Xenophon was not a philosopher. The Socrates of Plato is Xenophon's Socrates more or less transfigured, certainly much refined. In a certain sense and measure Socrates and Plato were, each of them, at once the creature and the creator of the other.

According to various more or less credible, though somewhat confused and not unquestioned, reports Plato devoted twelve years, following the death of Socrates, to traveling extensively in quest of knowledge. We are told that, after taking refuge for a time with the philosopher Euclid in Megara from possible violence against conspicuous admirers of Socrates, Plato visited Egypt, Persia, India, Cyrene, Phoenicia, Babylonia, Judea, Sicily, and Italy where, it is said, he joined a school or sect of Pythagoreans.

At the age of forty, having now completed his many far wanderings in search of wisdom, Plato returned to Athens and there established, in the grove of Academus, a school, thenceforth called the Academy, where he is believed to have taught during the remaining two score years of his life. How he conducted the school, what he taught there, and what was the manner of the teaching, are questions that cannot be answered except by conjecture. In course of the nine hundred years of its existence—it was closed by an edict of Justinian in 529 A.D.—the Academy had many an eminent man for its head but only one Plato, and many pupils destined to fame but only one Aristotle.[8]

Regarding Plato the philosopher, it has been said that he teaches you nothing but

---

8 Edward Gibbon: "Yet the golden chain, as it was fondly styled, of the Platonic succession, continued forty-four years from the death of Proclus [in 485 A.D.] to the edict of Justinian, which imposed a perpetual silence on the schools of Athens, and excited the grief and indignation of the few remaining votaries of Grecian science and superstition."

makes your eyes bright. Rightly understood, the saying is true. For the brightness that is meant is not the superficial brightness produced by reflection of solar light; its source is psychic light, it is a shining from within, kindled there by the magical discourse of the Son of Apollo, as Plato has been significantly called; and the "nothing" he teaches you is not a nothing of the things apprehended by the spirit but a nothing of the dull and shabby stock-in-trade of an uninspired, pedantic, visionless didacticism.

The list of the dialogues is impressive: the *Apology* of Socrates, *Crito*, *Phaedo*, *Republic*, *Gorgias*, *Symposium*, *Protagoras*, *Phaedrus*, *Parmenides*, *Statesman*, *Charmides*, *Lysis*, *Laches*, *Euthydemus*, *Cratylus*, *Ion*, *Meno*, *Euthyphro*, *Timaeus*, *Critias*, *Theaetetus*, *Sophist*, *Philebus*, and *Laws*. There are other dialogues which, though ascribed to Plato, are held by competent scholars to be of doubtful authenticity. Even the *Laws*, commonly regarded as the latest of Plato's works, is so lacking in the charm, verve, elevation, liberality, and other Platonic qualities that some have judged it to be the work of an inferior hand. Of all the dialogues the first ten in the foregoing list are commonly thought to be the most important, and the *Republic* is the greatest of the ten. Certainly they are among the most

"[B]eauty
only, absolute,
separate, simple,
which without
diminution and
without increase,
or any change,
is imparted to
the ever-growing
and perishing
beauties of all
other things."

PLATO,
SYMPOSIUM

precious things that have come down to us from the creative genius of antiquity. To neglect them is to neglect something essential to truly liberal culture. There is nothing to take their place.

Here it is impossible to analyze any one of them or even to characterize it briefly. Suffice it to say that the dialogues are dramas. What they dramatize is the life of reason bent on human emancipation, not only on freeing men from the grip and down-rending pull of whatever is vicious or mean or sordid in the world, but especially on finding and showing them how they may realize their inborn potentialities for making their individual and communal lives shining embodiments of the highest excellence—the virtue, that is, of intelligence, knowledge, and wisdom, all attuned to the harmony of the good, the true, and the beautiful. And so the central themes of the great discourses are, for the most part, themes of universal and everlasting interest: the Nature of the Soul, for example, Immortality, Justice, the Ideal City or State, the Nature and Service of Philosophy, the Reality of Ideas, Education, Friendship, Love, and Death.

The saying of Emerson that "Plato is philosophy, and philosophy Plato"[9] is no doubt an exaggeration. But it is no exaggeration to say that in the history of philosophy after 347 B.C. there is to be found neither insight nor utterance quite so divine as the vision and the voice of Plato.

---

9 Alfred North Whitehead's similar sentiment in his 1927-1928 Gifford Lectures: "The safest general characterization of the European philosophical tradition is that it consists of a series of footnotes to Plato. I do not mean the systematic scheme of thought which scholars have doubtfully extracted from his writings. I allude to the wealth of general ideas scattered through them. His personal endowments, his wide opportunities for experience at a great period of civilization, his inheritance of an intellectual tradition not yet stiffened by excessive systemization, have made his writing an inexhaustible mine of suggestion."

ARISTOTLE

3

ARISTOTLE

384-322 B.C.

ARISTOTLE—PERHAPS THE MOST stupendous, certainly the most influential, of philosophical thinkers, long honored as "the Philosopher" and "Master of those who know"—was born in 384 B.C. at Stagira, a Macedonian city about two hundred miles north of Athens. The place of his birth accounts for his being familiarly designated as the Stagirite. His mother, Phaestis, and his father, Nicomachus, physician to the king of Macedon, were both of them of distinguished ancestry.

The medical atmosphere in which he was brought up and the discipline which he as a physician's son probably received in the art of dissecting, in accord with the prevailing practice of his father's school of medicine, naturally favored the development of the youth's scientific predilections, and may account for the story that, when a very young man, he practiced medicine in Athens.

Upon his father's death, Aristotle, then eighteen, went to Athens to study and there entered Plato's Academy, where he remained twenty years. Neither before nor since has so great a teacher had so great a pupil, or so great a pupil so great a teacher: a conjunction of geniuses quite unmatched in the history of education. Plato was the elder by more than forty years. He spoke of Aristotle as the *Nous* (mind or intellect) of the school and referred to his home as "the house of the reader," for though Euripides was the first,[10] Aristotle was the second Athenian citizen to establish a large private library. Later when the pupil, seeming to have become perhaps a bit chesty, began to show signs of depreciating some of the Master's teachings, the old sage compared the youngster to

---

10 Euripedes, Greek dramatist, thought to have built the first library in Athens, possibly located in a cave with a view of the sea.

"The poet [Euripedes] lived, so Philochorus says, on his own estate at Salamis, and worked in a cave facing the sea . . . [He] was the first Greek who collected a library, the writer and thinker, not the man of affairs."

GILBERT MURRAY,

1897

a foal that kicks his mother after draining her dry.

It would be erroneous to imagine that during his twenty-year connection with the Academy Aristotle was a mere learner. During part of this time of his pupilship he was also a teacher and a writer. Though he may not have taught in the Academy, he at least had pupils in Athens, Eudemus of Cypres [Cyprus],[11] for example, Theadectes, and Hermias, who was later to have a court of his own as autocrat of Atarneus and to give his sister or niece to Aristotle for wife. It is even reported that during the period in question Aristotle founded and conducted a school of rhetoric to rival Isocrates of whom he disapproved. While still a pupil of Plato, Aristotle wrote books, some of them dialogues after the manner of the Master, and some of them didactic in the manner of his own maturer works. Of these earlier writings only fragments survive.

When Plato died (347 B.C. at the age of eighty), his nephew, Speusippus, and not Aristotle, became head of the Acad-

11 Eudemus of Cyprus may have been a friend, rather than a student, and was killed attempting to overthrow the tyrannical king of Syracuse. He is not Eudemus of Rhodes, who was certainly a student of Aristotle and the historian of geometry referenced by C.J.K. in the portrait of Pythagoras.

emy. Thereupon Aristotle, now a mature man of thirty-eight, highly educated and already reputed for his intellect, attainments, and writings, left Athens and for a period of twelve years engaged in a variety of activities—traveling, observing, teaching, reading, meditating, writing—all favorable to the maturing of his philosophy. He went first, by invitation, to the court of Hermias, his friend and former pupil; next resided for a time at Mitylene. When forty-three he was summoned by Philip of Macedon to teach his son Alexander, then thirteen, and served in this capacity for three years, without, however, taming the native savagery of the frightful youngster soon to be the ruthless conqueror of the world. He appears to have had other memorable pupils, Callisthenes, Cassander, Marsyas, Phanias, and Theophrastus.[12] A part of the twelve-year period he spent in his native city, Stagira, which he provided with a new constitution after it had been sacked by Philip and then restored by him or his son in response to Aristotle's appeal. Writings begun before leaving Athens were advanced, and time was given to the observational study of animals, for which the means, it is said, were generously provided by the Macedonian kings.

At the age of fifty (some say fifty-three), Aristotle returned to Athens—now smarting under Macedonian rule, which Aristotle, however, preferred to Athenian democracy—and there established in the Lyceum a school of his own, destined to last, with many changes of spirit and doctrine, for two hundred and seventy generations, till the Christian emperor Justinian closed it with the other pagan schools of Athens in 529 A.D. The school was soon dubbed Peripatetic (its members Peripatetics) either because its garden had a covered walk (*Peripatos*) or, as some say, because Aristotle lectured to his pupils while walking with them.

Aristotle's school was no copy of Plato's Academy. They differed profoundly in aim, in manner, in spirit, in range of interest, in curricula. Plato was a poet, a great artist, a dramatist of Reason, impassioned for human amelioration, and withal a good deal of a mystic. Aristotle was none of these things. He was cold, didactic, matter-of-fact, sober,

---

12 Callisthenes died in prison (and possibly was executed) for plotting against Alexander the Great. Cassander became king of Macedon.

restrained, without rapture or wing. His motto was *nil admirari*. Plato was interested primarily, almost exclusively, in the study of Man, not in that of inorganic things or of animals and plants. Aristotle, concerned with all these things, dealt with all of them both in his school and in his writings. Logic, Physics, Metaphysics, Ethics, Politics, Biology, all were embraced. If Aristotle had been born a few years earlier he might have been a pupil of Hippocrates.[13] It has been said that, if he had spent a year with him instead of twenty with Plato, he would have been a supremely great *scientist* (in the modern sense) instead of being what he was, primarily a metaphysician (in the ancient sense). Nevertheless, despite Plato's influence, he accomplished much in various sciences, especially in Biology, including Physiology and Anatomy. He practiced dissection, and in the first great zoological garden of the world, munificently provided for him by Alexander, he made many acute and important observations and classifications, coming perilously near to the conception of evolution.

In 323 B.C., Alexander died. Athens rejoiced, overwhelmed the Macedonian party, with which Aristotle was known to sympathize, and proclaimed its own independence. By a chief priest Aristotle was indicted for heresy, charged with having taught the vanity of prayer and sacrifice. Remembering the fate of Socrates, he decided to quit Athens, designated Theophrastus to be his successor in the Lyceum, and betook himself to Calchis where he died in 322 B.C. Aristotle's will, most of which has come down to us, shows clearly that the philosopher was very humane.[14]

Aristotle was a very prolific writer. Some ancient authors have credited him with several hundred volumes, some with a thousand or more. Such reports are silly exaggerations. Of all the volumes once ascribed to him many are known to be spurious, a few are still in doubt, but there remain enough works of unquestioned authenticity to seem a miracle of achievement. The chronological order of their production is uncertain. It is highly probable that at least some of them were

13 Probably a reference to Hippocrates of Kos, the Greek godfather of medicine. Hippocrates of Chios was an astronomer.

14 The will makes various provisions for Aristotle's children, friends, and servants, and the conditional emancipation of several slaves.

written simultaneously in the course of perhaps thirty-five years.

Plato *dramatized* the life of reason. Aristotle did not, he *analyzed* it. It goes without saying that, for those who are qualified to judge them, the *Metaphysics*, the *Politics*, and the *Ethics* are possessions forever.

Except his biological works, which are valuable despite the presence in them of colossal errors, Aristotle's so-called "scientific" works are today chiefly interesting, not as science, but as items in the history of speculation. Aristotle was not an experimenter, and his *Physics* is not the outcome of experiment, observation, hypothesis, and verification but chiefly of a *priori* reasoning—not physics but metaphysics. The same is true of the astronomy. In fairness we have, of course, to remember that he was without scientific instruments—telescope, microscope, clock, etc., to say nothing of mathematical analysis.

Perhaps the most valuable, certainly the solidest, of Aristotle's achievements are those works constituting what, not he, but his followers named *Logic*. Aristotle's *Logic*, after a reign of over two thousand years, is still in vogue. It is in the main deductive logic, the theory of the syllogism, not inductive. For many centuries it was regarded as both perfect and complete, complete in the sense of embodying all the processes of ratiocination, just as Euclid's

"Nothing is great to [a great man]."

ARISTOTLE,
NICOMACHEAN ETHICS

Elements was thought to be both perfect and complete, complete in the sense of containing the whole of possible geometry. We know now that Aristotle's *Logic* is only a small, though a precious, part of Logic, just as Euclid's *Elements* is only a small, though a precious, part of Geometry.

"Plato *dramatized* the life of reason. Aristotle did not, he *analyzed* it."

CASSIUS JACKSON KEYSER

Though the *Logic* of Aristotle held its own from the very beginning, Aristotelianism as a philosophic *system* did not. The system as such soon lost its vogue and by the sixth century A.D. had been forgotten. But it came to memory again, and thereafter, by means of translations into other tongues, first into Syriac, thence into Arabic, thence, later, into Latin, it gave inspiration and guidance to such Arabic scientific and philosophic thinkers as Avicenna and Averroes,[15] finally succeeded in dominating the scholasticism of the Middle Ages, and thus exercised an almost undisputed sway in cultural centers East and West for a thousand years.

15 Avicenna Ibn Sina, the Persian philosopher, refers to Aristotle as the "First Teacher." Averroes Ibn Rushd, an Andalusian philosopher agrees: "[T]he teaching of Aristotle is the supreme truth, because his mind was the final expression of the human mind."

EPICURUS

4

EPICURUS

341-270 B.C.

EPICURUS—TODAY THE MOST FAMOUS, though hardly the most original, of the Greek Atomists—was born at Samos near the beginning of 341 B.C. His father, Neocles, seems to have been a school teacher. Of the son's education almost nothing is known. At the age of eighteen, Epicurus went to Athens. The great school of Plato, who had been dead twenty-five years, was flourishing under the leadership of Xenocrates. Aristotle, accused of impiety and remembering the fate of Socrates, had just made Theophrastus head of the Lyceum and withdrawn to Chalcis, where he died the next year, 322 B.C. In this year Epicurus rejoined his father, now at Colophon. Here, and at Mitylene and Lampsacas, he lived about fifteen years, probably teaching, certainly philosophizing and acquiring devoted disciples.

At the age of thirty-four he returned to Athens, bought a garden for about $1,500 [80 minae or 8,000 drachma]—destined to fame as the "Garden of Epicurus"—and there established a new type of community, composed of men and women devoted to his philosophical principles, a kind of close brotherhood, so to say, of his disciples. Of this unique society of friends—distinguished for their plain living and high thinking, for their devotion to study, friendly discussion, and writing, and for their hearty cooperation in promoting the development and understanding of their common philosophy—Epicurus was the beloved head and leader, guide, counselor, and friend, for thirty-six years, till his death at the age of seventy. It does not appear that in this school there was any lecturing or other formal instruction as in rival schools, though there was much "learning by heart." It goes without saying that this singular community was often misunderstood and much lied against. Nevertheless the city erected statues in the founder's honor and it is known that his friends "were many enough to fill whole cities."

Epicurus is reputed to have been the most prolific writer among the thinkers of ancient Greece. The rolls of his books

"[If] Epicurus was at pains to keep his pupils strictly to the letter of his doctrines, he succeeded so well that none of them is known to have made any attempt worth mentioning towards their development."

EDUARD ZELLER,
1883

are as many as three hundred. The titles of the books are known. But of the books themselves nothing remains save four handbooks (abridgments of his teachings), which were fortunately preserved by Diogenes Laërtius. Nevertheless, the philosophy of Epicurus is well known.

Epicurus was a remarkable union of saint and scientist. As a saint his dominant desire was to teach the art of so living as to secure the fullest measure of well-being consistent with the Nature of Things. As a scientist he aimed at discovering and teaching what the Nature of Things really is.

In attempting to estimate the endeavors and achievements of Epicurus it is essential to reflect that he belonged to a scientifically young generation. In comparison with him, his contemporaries and predecessors, it is we of today who are the ancients, not they; and no doubt the time will come when we, who are now the ancients, will be regarded as having belonged to the youth of scientific Man. To ignore that fact and scoff is to be, not a philosopher or historian or critic, but a fool. Epicureanism is not dead. Many of its crudities and blunders have "perished by supersession"[16] but its spirit survives, for it is the spirit of Science trying to understand the World and Man.

Epicurean thought embraces three grand divisions: the Physics, presenting a cosmology; the Canonicon, dealing with epistemology and scientific method; and the Ethics, at once the final cause and the crown of the system as a whole.

In the Physics, Epicurus set himself a problem never surpassed in scope or in audacity. It is the problem of constructing a cosmology according to which all things in Nature—all processes, all events, all facts and phenomena, material or mental, physical or psychical—could be so explained that the Universe would no longer appear to be a created, transitory, chaotic affair, theater and playroom of intervening, irresponsible, capricious gods, but would be seen to be a genuine Cosmos, uncreated, lawful, humanly intelligible, and everlasting.

According to his solution of the problem, the Universe of things, organic or inorganic, bodies or minds, is wholly composed of three and but three things: Matter, Motion (of matter), and infinite Space (including Void, empty space). Matter is composed of an infinity of imperceptibly minute, absolutely solid, non-decomposable, perpetually moving, and indestructible atoms, "the first beginnings" or "seeds of things." The atoms differ in shape; of shapes the number, though large, is finite; but the number of atoms of a given shape is infinite. All transitory things, organic or inorganic, large or small, bodies or minds, worlds or pebbles, have been produced by the clashing together and interlocking of atoms, and will in due course be again reduced to atoms by the hammering of atomic storms perpetually beating against all existing things. So it always has been and always will be.

According to this austere doctrine cosmic history is literally a beginningless and unending succession of unpurposed generations and destructions—an infinite and eternal drama of Venus and Death.[17] Of that bold conception

---

16 C.J.K. praising *De Rerum Natura*: "I maintain that opinion of the work despite the fact that the majority of its conclusions have been invalidated by time, have *perished by supersession*; for we must not forget that, in respect of knowledge, 'the present is no more exempt from the sneer of the future than the past has been.'"

17 Venus was the Roman goddess of fertility. Mors is the Roman Death.

Epicurus professed to be the sole originator, but it is known that he derived much of it directly from the great Democritus, some of whose now lost works he had read, and indirectly from Leucippus, the founder, according to Aristotle and Theophrastus, of Atomism.

But Epicurus was far from being a mere imitator or copyist. There is room here to mention only two of the important differences between him and Democritus. One relates to the question of determinism and freedom of will. Democritus was a thoroughgoing determinist. But Epicurus provided for freedom by postulating among the bases of his philosophy a principle of indetermination. As given by Lucretius the postulation, with its motive, runs as follows:

> When bodies are borne downwards sheer through void by their own weights, at quite uncertain times and intervals they push themselves a little from their course: you just and only just can call it a change of inclination...If all motion is always one long chain, and new motion always arises out of the old in order invariable, and if the first-beginnings do not make by swerving a beginning of motion such as to break the decrees of fate, that cause may not follow cause to infinity, whence comes this free will in living creatures all over the earth, whence I say is this will wrested from fate by which we proceed whither we please, swerving also our motions not at fixed times and fixed places, but just where our mind has taken us?

Again, Democritus was a mathematician. Among his seventy works, now lost, twelve were mathematical. Epicurus, however, was not only no mathematician, but mathematics was one of the many cultures that he despised.[18] Nevertheless the history of mathematical ideas must reckon with what he said of the importance of distinguishing between the finite and the infinite.

The cardinal doctrine of the Canonicon is that knowledge of reality depends initially and finally on Sensation and

---

18 According to Sextus Empicirus, Epicurus sought to "make war against the Arts and Sciences."

Feeling. Conception and Inference, or Reason, are also indispensable but sensations are the source of conceptions and the ultimate test of inferences.

The *Ethics* bases itself on the postulate that Pleasure is the only good. But, said Epicurus:

> It is impossible to live a pleasant life without living wisely and well and justly....

Of the philosophic, or scientific, world-systems bequeathed to us by the creative genius of antiquity, that of Epicurus is, in at least one respect, the greatest. There is no other in which the historian of thought can find so many of the ancient roots of modern science. For one there encounters such concepts and dogmas as: natural law, the denial of divine interference, the atomic constitution of matter, conservation of mass and of energy, organic evolution, evolution of human society and civilization, origination of life from the inorganic, spontaneous generation, struggle for existence, survival of the fit, destruction of the ill adapted, heredity, the puzzling nature of time, logical induction, the rejection of teleology, the perpetual motion of particles seemingly at rest, the continual birth and destruction of worlds, sensation and reasoning as the sole means to knowledge.

In the second century A.D., the philosopher and Roman emperor Marcus Aurelius endowed chairs of philosophy equally in four then already ancient Athenian schools—the Platonist, the Peripatetic, the Stoic, and the Epicurean.[19] The school founded by Epicurus lasted as an institution for over six hundred years, and its influence, clearly manifest in the Renaissance, is still felt throughout the world.

---

19 Salary for a philosophy chair: 10,000 drachmas, a bit more than the purchase price of the Garden.

ROGER BACON

5

ROGER BACON

ca. 1214-1292

Roger Bacon, Albertus Magnus [St. Albert the Great], St. Bonaventura, and St. Thomas Aquinas—contemporaries, personally known to one another—were doubtless the greatest four among the summit-minds of the great thirteenth or other medieval century. Of the four the three scholastics—Albertus, Bonaventura, and Thomas—achieved high fame in their own day and beheld the glory of their own renown. But Bacon did not. On the contrary he was but little appreciated, much persecuted and harassed, long officially restrained from writing anything for publication, and, at length, when an old man, was condemned on the charge of teaching "suspected novelties" (not further specified) and imprisoned for fourteen years. The intervening centuries have greatly altered the perspective. So much so that today Roger Bacon is generally, though not universally, regarded not only as having been what Alexander [von] Humboldt called him, "the greatest Apparition of the Middle Ages," but certainly by a great majority of competent critics, as having been a flaming scientific genius born four hundred years before time was ripe for such a birth.

Of the outward facts of Bacon's life very little is known, and most of this little is derived, mainly by inference, from allusions to himself and others in his writings.

He was born about 1214 at or near Ilchester, England; of his death the most probable date is 1292; he was buried in the Franciscan Church at Oxford.

Thus his life fell at a time when the social order was of the feudal type. There was the State hierarchy—God, King, Nobles, and so on down. And there was the Church hierarchy—God, Pope, Cardinals, Bishops—in descending order down to the undistinguished mass of the people. Every man was subject to, dependent on, a "superior." To be "lordless" was to be an outlaw.

Both within and without the universities the supreme intellectual interest was Theology. Science and its method were virtually non-existent. "Truth" was discovered, established, handed down, dictated, imposed by infallible Authority—the Bible, decrees of Popes and Councils, writings of the Christian Fathers, and, strangely enough, the pagan philosophy of Aristotle.

Bacon's family, said to have been noble, was well-to-do, perhaps wealthy. Roger speaks of an elder brother as "rich," having inherited the family estate, and as having been subsequently impoverished and exiled for his support of Henry III in the Barons' War.[20] Another brother is mentioned as a "student" or "scholar."

Roger studied at Oxford and later at the University of Paris. Having a strong scientific bent and love for secular learning, he had the great good fortune at Oxford to have at least one eminent teacher suited to his predilections, Robert Grosseteste, then Chancellor of Oxford, and later Bishop of Lincoln. For this man, though primarily a theologian, had a lively, if only secondary, interest in "natural" knowledge, founded the study of Greek at Oxford, and was, said Bacon, the only man there who knew mathematics, optics, and the other sciences, all dear to young Bacon's heart.

Bacon himself became a teacher at Oxford and subsequently a member of the Paris faculty. His interests and activities as student and teacher are revealed in his own words written at the age of about fifty-three:

> I have labored from my youth
> in the sciences and languages,
> and for the furtherance of study,
> getting together much that is
> useful. I sought the friendship
> of all wise men among the
> Latins, and caused youth to be
> instructed in languages, and
> geometric figures, in numbers
> and tables and instruments,
> and many needful matters. I
> examined everything useful to
> the purpose, and I know how to
> proceed, and with what means,
> and what are the impediments:
> but I cannot go on for lack of
> the funds which are needed.
> Through the twenty years in
> which I labored specially in the
> study of wisdom, careless of the
> crowd's opinion, I spent more
> than two thousand pounds in

---

20 The First Barons' War: nobles back a French prince as their king, then shift course and back Henry III.

> these pursuits on occult books and various experiments, and languages and instruments, and tables and other things.

After devoting twenty years to scientific research, in which he spent his fortune, having sometimes to send overseas for specimens and instruments, Bacon became a Franciscan about 1247 during his sojourn in Paris. Why he joined the Order is unknown. After returning to Oxford he incurred the Order's suspicion, and was sent back to Paris where his freedom of expression was officially restrained. He later refers to himself as having been "for ten years an exile."

Prior to 1266, Bacon seems to have written only one work of importance, *De Multiplicatione Specierum*. But in that year there came a great crisis in his life. Pope Clement IV, who had become interested in Bacon before being made Pope, sent "to our beloved son, Brother Roger, called Bacon," a letter commanding Roger to send him without delay copies of all his works, "notwithstanding the prohibition of any prelate, or any constitution of thy Order." Naturally Bacon was ineffably elated, for here, it seemed, was a chance to win for scientific research and learning the invincible support of the Head of Christendom. Unfortunately His

"Now,
like Friar Bacon's
brazen head,
I've spoken.
Time was, Time is,
Time's past."

LORD BYRON,
DON JUAN

Holiness erroneously believed the works with which Bacon's head and heart were throbbing had been reduced to writing. Bacon had long before projected an encyclopedic work to comprise four large volumes dealing, respectively, with Grammar and Logic, with Mathematics, with Physics and Natural Philosophy, and with Metaphysics and Morals, including all their many subdivisions and ramifications. To compose this immense work in response to the Pope's letter Bacon found impracticable and so he compromised by writing, in about one year, and sending to him three preliminary or introductory reports, the *Opus Majus*, the *Opus Minus*, and the *Opus Tertium*. From Clement, who died the following year, Bacon received no response.

Master of the art of caustic speech, Bacon sometimes "attacked dead and living worthies tactlessly, fatuously, and unfairly." But this does not imply that his bitter and courageous arraignment of the ignorance and corruption of high-placed men regarding themselves, and stupidly regarded by others, as "authorities" in matters scholastic and ecclesiastic was essentially unjust.

Bacon's fourteen-year imprisonment for teaching "suspected novelties" began about 1277 and ended only shortly before his death in 1292.

In recent years, Bacon literature has become extensive, some of it more or less derogatory but most of it highly laudatory. What have such experts independently said of Bacon? *Doctissimus mathematicus*; outliner of a true science of agriculture; worthy of a place among the chemists; the originator of experimental research; discoverer of gunpowder; independent thinker in the philosophy of languages; in medicine a giant; owner and probably inventor of a telescope and compound microscope: such are specimens of the estimates.

Bacon's brilliant dreams of the flying machine, high-speed engine, self-propelled ships, and other marvels, are familiar.

His understanding of the scientific method is evident, not only in his deeds, but in his words:

> There are two modes of acquiring knowledge—by reasoning and by experience. Reasoning draws a conclusion, and makes us grant the conclusion, but does not make the conclusion certain, nor does it remove doubt so that the mind may rest on the intuition of truth, unless the mind discovers it by the path of experience...He who

> wishes to rejoice without doubt in regard to the truth's underlying phenomena must know how to devote himself to experiment...You could never by mere argument convince a man that fire would burn. In spite of the demonstration of the properties of an equilateral triangle the mind would not stick to the conclusion without experience.

And Bacon envisaged progress like a modern:

> The study of wisdom may always increase in this life, because nothing is perfect in human discoveries. Therefore, we later men ought to supplement the defects of the ancients, since we have entered into their labors, through which, unless we are asses, we may be incited to improve upon them. It is most wretched always to be using what has been attained, and never reach further for one's self.

Was Bacon a heretic? He was no more a heretic than was Descartes or Newton or Leibniz. Even a great personality may lack self-consistence. Theologically Bacon belonged to his age, but scientifically he was centuries in advance of it. Had Roger Bacon been born four hundred years later, he would have been the peer and fit companion, perhaps the leader, of the seventeenth century's great pioneers of Modern Science.

RENÉ DESCARTES

6

RENÉ DESCARTES

1596-1650

René Descartes, mathematician, scientist, and philosopher, was born at La Haye, France. His parents were gentle folk, neither poor nor very rich. The mother, Jeanne Brochard, died very soon after René's birth. The father, Joachim Descartes, lawyer and member of the provincial parliament, married again but remained devoted to his "little philosopher."

Though not specially precocious, the boy was inquisitive, quick to learn, and "ardently desirous of instruction." But because his health was frail, his formal education was not begun till the age of eight, when he entered the Jesuit college of La Flèche, then one of the most celebrated schools in Europe. During his eight years here he not only mastered readily the entire curriculum of languages, history and fable, logic and mathematics, eloquence and poesy, ethics, theology, and philosophy, but in addition read, on his own account, many rare and curious books, all he could get, on extra-curricular subjects. Here, too, he made many friends, two of them notable: the rector, Father Chalet, owing to whose kindly encouragement he formed the lifelong habit of remaining in bed of mornings till eleven o'clock; and a schoolmate, Marin Mersenne,[21] who later became Father Mersenne and for many years served Descartes as his literary and scientific agent and as guardian of his privacy.

For mathematics, especially for its method, he acquired in college a respect both deep and abiding, but not so for the other subjects, and on quitting the school he entirely abandoned the study of letters as affording neither genuine knowledge nor an effective method of research. With a view to gaining knowledge of himself

---

21 Mersenne served as moderator of a proto-listserv, disseminating scientific discoveries around Europe by handwritten letter.

"I only desire tranquility and rest, which are goods that the most powerful Kings of the Earth cannot give to those who cannot take of themselves."

RENÉ DESCARTES,
1650

and the world he next devoted thirteen years to meditative traveling, visiting courts and armies, serving as a soldier in various wars, observing customs and institutions, holding intercourse with men of different dispositions and ranks including distinguished mathematicians, and, most notable of all, secluding himself from time to time in various places for prolonged and undisturbed reflection upon the experience thus gained.

As this period drew to its close, Descartes discovered that, though he had published nothing, he nevertheless had, doubtless owing to his numerous conversations and much correspondence with distinguished men, acquired a reputation for a wisdom which he knew he did not possess, for his philosophic enquiries had been barely begun. Astonished and somewhat chagrined by that discrepancy between his reputation and his achievement, he resolved to redress the balance by henceforth devoting himself wholly and systematically to scientific and philosophic research.

The "tranquility and repose," which were for him essential to the enterprise, he sought and found in Holland, where he lived twenty years (1629-1649) and produced his famous works. The annals of human thinking contain nothing to surpass the grandeur of Descartes' philosophic aim. *The Discourse on Method*, greatest of his works, is a charming tale of his own mental life in seeking, finding, and practicing the method by which, as he was convinced, it was possible to establish a universal science, a thoroughly rational cosmology, a true theory and understanding of the Universe and Man.

Rightly to estimate Descartes as a mathematician it is necessary to consider two very distinct things: (1) his contribution to mathematics as such; (2) the role of mathematics in fashioning his philosophic method. As to the former the publication of his *La Géométrie*, founding what is now known as Analytic Geometry, must be reckoned among the greatest events in the history of science and mathematics, for that single work did more than any other to emancipate the mathematical spirit from the two thousand years' dominion of Euclid and thereby made possible the subsequent great mathematical developments that now constitute what is perhaps the chief intellectual glory of the modern world.

As to the second point, the influence of mathematics on Descartes' philosophic methods, his own words are revealing:

> The long chains of simple and easy reasonings by means of which geometers are accustomed to reach the conclusions of their most difficult demonstrations, had led me to imagine that all things, to the knowledge of which man is competent, are mutually connected in the same way, and that there is nothing so far removed from us as to be beyond our reach, or so hidden that we cannot discover it, provided only we abstain from accepting the false for the true, and always preserve in our thoughts the order necessary for the deduction of one truth from another.

Thus did mathematics furnish Descartes with the ideal which his all-embracing cosmology, he hoped, was closely to approximate or even to realize, so that his aim and dream of a universal science were to be in fact a universal mathematics. The astounding task's demands were obviously two: (1) a set of indubitable truths, to serve as principles; and (2) deduction from these, in mathematical fashion, of the true doctrine of the *World*.

Descartes sought to meet the first demand by a long and arduous self-discipline in skepticism, rejecting all dubitable propositions, holding for true only such as he was unable to doubt. The most

"Descartes was a philosopher, not a physiologist. He took interest in the problems of the living body only so far as they bore on the greater problems of the why and the wherefore of the universe."

MICHAEL FOSTER, 1924

basic of the indubitables thus found is the famous "I think, therefore I am." Upon a few such principles he erected an imposing cosmology which, though now superseded in many important respects, will remain an eternal witness to the versatile creative genius of its author. But, not being a logically concatenated system, it is not a universal science or a universal mathematics except perhaps in its intention, its spirit, and its sweep.

As to the mental climate of Descartes' time, it is enlightening to recall why his earliest work, the *World*, setting forth his physical theory of the universe, holding, in particular, that the universe is infinite and that the earth revolves around the sun, was not published till after the author's death. About to send the work to press by his friend Mersenne in 1633, he was deterred from doing so by news of what had just happened to another man of science. Descartes writes to Mersenne: "But I have just been at Leyden and Amsterdam to ask after Galileo's cosmical system as I imagined I had heard of its being printed last year in Italy. I was told that it had been printed, but that every copy had been at the same time burnt at Rome, and that Galileo had been himself condemned to some penalty." He probably recalled, too, the sad fates of [Giordano] Bruno, [Petrus] Ramus, [Lucilio]

Vanini, and [Tommaso] Campanella.[22] At all events, not aspiring to the glory of martyrdom, he consigned the manuscript of the *World* to his desk.

Nevertheless, once the *World* was published, its very ingenious doctrine of vortices enjoyed an almost undisputed scientific vogue for a score of years, thereafter yielding but slowly to its rivals.

In the early seventeenth century, what we today call the scientific method, then struggling into existence, was understood by only a few. It has been denied that Descartes was one of these; that in religion he depended on authority, which is true; that in science he often resorted to *a priori* reasoning, which is also true; and that he failed to appreciate the importance of observation and experiment, which is not true; for by observation and experiment he established weighty results not only in physics but in anatomy and physiology (which in his view were branches of physics). Sir Michael Foster's assertion that Descartes was no physiologist but merely wrote on the subject speculatively is very

"Descartes is a type of that spirit of science to which erudition and all the heritage of the past seem but elegant trifling."

WILLIAM WALLACE, 1910

22 A somewhat eclectic list of unrelated philosophers caught up in heresy, religious warfare, and political intrigue. All were either executed, assassinated, or died in prison.

unjust as shown by Thomas H. Huxley[23] in his lecture on Animal Automatism (a Cartesian doctrine) where Huxley says that Descartes was "entitled to the rank of a great and original physiologist...an unwearied dissector and observer" and quotes from Descartes a series of propositions "constitut[ing] the foundation and essence of the modern physiology of the nervous system." No great reader in his mature years, Descartes was a great and independent investigator who regarded mere erudition as but "elegant trifling," and who, on being asked by a visitor to see his library, pointed to the animals in his dissecting room and said, "Here are my books." Regarding the importance of experimentation he wrote poignantly:

> I am becoming daily more and more alive to the delay which my design of self-instruction suffers, for want of the infinity of experiments I require, and which it is impossible for me to make without the assistance of others.

In 1649, Descartes very reluctantly accepted a pressing invitation from the nineteen-year-old queen Christina to the Swedish court. Here, obliged despite the rigors of winter to go daily to a chill room of the palace to give the queen a lesson at five in the morning, he contracted inflammation of the lungs and on February eleventh died at the age of fifty-four, a victim of the inordinate vanity of a presumptuous girl.[24]

---

23 Grandfather of Aldous Huxley, *Brave New World*. Incidentally, the same Sir Michael Foster edited Thomas Huxley's memoirs.

24 Queen Christina abdicated the throne and converted to Catholicism. Among her supporters was Leibniz.

BLAISE PASCAL

7

BLAISE PASCAL

1623-1662

Blaise Pascal, the most puzzling personality in the history of European thought, was born at Clermont, Auvergne, France. His mother, Antoinette Begon, died when he was four. His father, Étienne Pascal, nobleman and lawyer, held a post in the civil service, as had the Pascal family for many generations; keenly interested in geometry and the study of Nature, he was a friend and associate of eminent scientists and mathematicians.

Blaise had two beautiful and talented sisters, Gilbert[e] and Jacqueline, the former older, the latter younger, than he. Jacqueline, a poet in her youth, later renounced the world and entered Port Royal. Gilbert[e], who became Madame Périer, was deeply devoted to her brother, as was Jacqueline also, and after his death wrote a charming account of his life, the famous *La Vie de Monsieur Pascal*.

Upon the mother's death, the education of the children, especially that of the son, became the father's dominant concern, and he soon moved with his family to Paris; but not for the schools. For he knew that these stupidly belabored the heads of eight-year-olds with the bones of Latin grammar; he knew, too, that school and college alike maintained towards natural science the attitude of ignorant hostility. Hence the boy was sent to no school nor did he ever attend a college.

Blaise Pascal was educated by his father, his sole teacher and, it seems, his only male companion. The pupil was a very precocious lad. "People," said Leonardo da Vinci [actually Niccolò Machiavelli], "fall into three classes, those who do not see, those who see when shown, and those who see by themselves." Young Pascal, who belonged to the third class, had not only remarkable quickness and keenness of observation but also the ability to ask significant questions, to demand rational answers, and, when these were not forthcoming, to find them himself by resort to experimentation. For example, impressed and puzzled by the behavior of sound he made many experiments and wrote a solid little treatise on the subject at the age of twelve.

Fortunately his precocity did not lead to the forcing of his education. On the contrary its pace was deliberately held in check. Latin and Greek were not begun until the boy was twelve when his father had already given him, by means of many conversations, much insight into the general nature of language. Mathematics, which the youngster eagerly enquired about, he was absolutely forbidden to study or to think of or even to mention until he should secure a good knowledge of the classical languages. Mathematical books were actually locked up and Étienne forbade his scientific friends to speak of mathematics in the boy's presence.

It would perhaps have been no great hazard to wager that this youth, if favored by health and length of years, would eventually accomplish something memorable. But no one could by any chance have foreseen that, despite an obscure malady making much of his life increasingly miserable and ending it at the age of thirty-nine, he was yet destined by the miracle of his genius to achieve, not merely some high distinction, but a seven-fold immortality—immortality as a mathematician, as a physicist, as an inventor, as the chief creator of his nation's great prose, as a theologian, as a philosopher, and as a fanatic.

According to Gilbert[e] the lad's first mathematical feat, probably unmatched in the story of mathematical prodigies, was his original discovery of the distinctive *method of mathematics*. For when he was thirteen years old, absolutely untutored in geometry, not even knowing its elementary terms but calling a circle a *rond*, a straight line a *barre*, and so on, his father one day surprised him in his playroom in the act of proving [Proposition 32 of Book I in Euclid's Elements] (respecting the sum of the angles of a triangle)[25] after having proved the logically prior propositions, all on the basis of axioms and definitions formulated by himself.

Thereupon he was given a copy of Euclid's *Elements*, which he quickly devoured and digested without difficulty or help. As a reward he was made a member of the *Académie libre*—a club

25 Euclid: "In any triangle, if one of the sides be produced, the exterior angle is equal to the two interior and opposite angles, and the three interior angles of the triangle are equal to two right angles."

composed of his father and his father's mathematical and scientific friends, the real beginning of what later became (1666) officially the French Academy of Sciences. Regularly attending the club's weekly meetings, the lad distinguished himself by the proposing and solving of problems and by helping in the appraisal of communications received from distinguished men of other countries.

Nevertheless, the members could scarcely believe their eyes when they saw the little *Essay on Conics*, which he published at the age of sixteen. Here occurs the now famous proposition that was destined to become central in modern projective geometry and is everywhere referred to as Pascal's Theorem. About the same time he wrote a *Complete Treatise on Conics*, seen and praised by Leibniz but, like several other mathematical writings of Pascal, was never published and is now lost.

Three hundred years ago the mathematical theory of probability (or chance), today so immense and vital, potentially infinite, and so potent or even indispensable as an instrument of research in logic, in philosophy, and in many branches of natural science, was not in existence, not even in embryo. It is the joint creation of two men either of whom could have done it alone, Blaise Pascal and Pierre Fermat,

"[His sister] relates that he discovered independently most of the first book of Euclid, that his intuition in mathematics seemed miraculous, and that geometry was simply his recreation."

DAVID EUGENE SMITH,
1923

and was begun in response to questions submitted to Pascal by a professional gambler. Essentially connected with his probability work is Pascal's *Treatise on the Arithmetical Triangle*, pioneering in Combinatorial Analysis.

The ablest mathematicians of the time wrestled much with certain problems of the Cycloid, called, because so beautiful, the "Helen of Geometry."[26] In his last burst of mathematical activity, driven thereto by insomnia after he had renounced mathematics and the world for religious meditation, Pascal conquered the problems, far surpassing all rivals, and did so by essentially employing the infinitesimal calculus (prior to its invention by Newton and Leibniz).

From Aristotle to the middle of the seventeenth century, physics taught the impossibility of a void (empty space). Instigated by [Evangelista] Torricelli's work Pascal, at a time when almost all preferred authority to reason and observation, proved by experimentation that the old doctrine is false unless every seeming vacuum is filled by some imperceptible and unknown matter like the unproved ether of [Christiaan] Huygens. No one has better described nor better practiced what we today call the scientific method. Pascal's hydrostatics, *The Treatise on the Equilibrium of Fluids*, is a scientific model, in content complete, in form perfect.

It is noteworthy that to this versatile genius the world is indebted for the invention of the first calculating machine and the omnibus.

Was Pascal a philosopher? No, if a philosopher is a system-builder, like Aristotle or Descartes, for, like Plato, Pascal built no system. But he was a great philosopher if a philosopher is one who thinks profoundly about the Universe and Man. For in respect of such thinking, whether we regard its amplitude or its clarity and rigor or its analytic penetration or the wealth of its intuition or its poignant eloquence or many another merit, the *Provincial Letters* and the *Pensées* of Pascal are hardly surpassed. In these, written after his abandonment of science, mathematics, and the world for the life of a religious solitary, are found Pascal's philosophy and theology fused into one.

The *Provincials*, "a triumph," wrote George Saintsbury, "of literary art of

26 Helen of Troy sparked the Trojan War. To draw a cycloid, glue a pencil to a wheel and roll the wheel along a wall. The curve that the pencil traces on the wall is a cycloid.

which no familiarity dims the splendour, and which no lapse of time can [ever] impair," written in behalf of Jansenism, is the most memorable episode in the long and bitter controversy between the Jansenists and the Jesuits.[27]

The *Pensées*, published after Pascal's death, is an immense chaotic ensemble of "Thoughts" jotted down in the recluse's last years in preparation for a magnificent *Apology for Christianity*, projected but never constructed.

No delineation of Pascal's philosophy is possible here. Suffice it to mention but one cardinal thesis, concerning Truth, Reason, and Grace. It is that mathematical truths are demonstrable by Reason; that truths regarding phenomena of external Nature can be discovered and established by observation and Reason; but that the truths of religion are above Reason's domain and can be known only by the Grace of divine revelation.

In his later years Pascal's habits became extremely ascetic. His final days were marked by violent spasms of physical pain. Postmortem examination revealed a serious lesion of the brain and serious disorders of the stomach and other organs.

27 It is difficult to disentangle the theological from the political elements of this controversy. That Jansen himself was dead before the controversy arose does not help matters. The most ostentatiously heretical tenet of Jansenism was that Christ did not die for all humanity. This was bound up in questions about predestination, free will, and even absolute monarchy.

BENEDICT SPINOZA

8

BENEDICT SPINOZA

1632-1677

In 1880 [actually 1877], two centuries after the death of Spinoza, a statue of the great philosopher was erected at The Hague by the means of an international subscription representing every part of the civilized world. On the occasion of the unveiling, the principal orator was Joseph Ernest Renan, who said in his peroration:

> Woe to him who in passing should hurl an insult at this gentle and pensive head.... From his granite pedestal this man will show to all men the way of blessedness which he found; and ages hence, the cultivated traveler, passing by this spot, will say in his heart, "The truest vision ever had of God came, perhaps, here."

Neither the monument nor the encomium had at any time been possible or even conceivable in the first of the two centuries, so universal and bitter was the odium that for a full hundred years clung to the great man's memory after his death and the quickly following publication of his *Opera Posthuma*. "People spoke of Spinoza," said Lessing, "as if he were a dead dog."

Why such abhorrence? The cause was, in sum, just the assertion of a mighty proposition—the Universe (or Nature) is God, and God is the Universe (or Nature)—a thesis rigorously maintained by Spinoza in the greatest of his works, the *Ethics*.

About 1780, after the lapse of a century, the abhorrence began to subside. A few intellectually emancipated men of great ability and candor—critics, poets, and philosophers—had begun to examine honestly and to ponder what the long hated thinker had actually written, and by the estimations of such as Lessing, Goethe, Herder, Schleiermacher, Novalis, Hegel, and Schelling,[28] "the holy and excommunicated Spinoza"[29] came to be at length recognized as one among the very greatest of the world's contributors to the Wisdom of mankind. Thought had indeed advanced. Less than two hundred years after Spinoza had incurred universal execration by his thesis concerning the identity of Nature and God, a great poet of Nature [William Wordsworth] could win the world's applause by singing of

> [S]omething far more deeply
> interfused
> Whose dwelling is the light of
> setting suns,
> And the round ocean, and the
> living air,
> And the blue sky, and in the
> mind of man;—
> A motion and a spirit, which
> impels
> All thinking things, all objects
> of all thought,
> And rolls through all things.[30]

On the Burgwal at Amsterdam is still shown the house where Benedict Spinoza was born in 1632. At this time Amsterdam contained a Jewish colony composed of families that had recently fled from Spain or Portugal to escape the cruelties of the Inquisition and had been attracted to the Dutch city by the assurance that there "no one would be molested on the account of his religion." The Spinozas, who were Jews, possibly Portuguese, probably Spanish, were members of that community. Benedict's grandfather had been for a time the community's head, and his father, a fairly prosperous tradesman, repeatedly served as warden of the synagogue. At the age of six, Benedict lost his mother; when twenty-two, his father.

Spinoza's earlier education was exclusively Jewish. Among his teachers were two men of distinction, Mannasse ben Israel and Saul Levi Morteira. Under them young Benedict plunged into the study of the Hebrew Bible, the Talmud, and the literature of Jewish medieval thought, which was traditional theology

---

28 All artistic or philosophical German writers, though only Johann Wolfgan von Goethe and Georg Wilhelm Friedrich Hegel have enduring fame in the Anglosphere.

29 "Heiligen verstoßenen Spinoza."

30 William Wordsworth, "Tintern Abbey," 1798.

tinged with Aristotelian and Neoplatonic elements. The keen-witted youth thus finding many problems but few solutions, began to seek light elsewhere. There was the Gentile world, but the language of Gentile scholarship was Latin. At this time, however, indeed during the seventeenth and eighteenth centuries, the Jewish schools made no provision for the study of Greek or Latin. From a certain [Franciscus] van den Ende—an accomplished man, intellectually independent, interested in science, educated as a physician but supporting his family by means of a private school—Spinoza learned Latin well enough to write it with facility and precision.[31] Thus was opened to him a vast new world, the world of Copernicus, of [Giordano] Bruno, and especially of Descartes whose teachings were then in great vogue.

Meanwhile, in accord with Jewish custom, he had learned a handicraft, the grinding of lenses for optical instruments, a skill that later brought him into touch with two of his great contemporaries, [Christiaan] Huygens and Leibniz, and led to his writing an optical *Treatise on the Rainbow.*

"[We] excommunicate, expel, curse and damn Baruch de Espinoza."

AMSTERDAM SYNAGOGUE, 1656

31 Van den Ende was executed in 1674 for conspiring to overthrow King Louis XIV of France.

"God,
or substance,
consisting of
infinite attributes,
of which each
expresses eternal
and infinite
essentiality,
necessarily
exists."

SPINOZA,
ETHICS, PART I,
PROPOSITION XI

Spinoza's studies had gradually given him certain firm convictions incompatible with the teachings of the synagogue. And, unwilling to seem what he was not, he ceased to attend its meetings. Neither official remonstrance, nor entreaties nor flattery nor warnings nor the secret offering of a bribe nor the attempt of a zealous ruffian to slay him with a dagger availed to restore his allegiance.[32] And so at the age of twenty-four the young philosopher was excommunicated—completely cut off from the commonwealth of Israel. For sheer malignity the curses officially invoked upon him are hardly surpassed in the annals of ecclesiastic malediction. The remaining twenty years of his life were devoted to solitary study, meditation, and writing; first, near Amsterdam, in the attic of the little house of a Mennonite friend whose simple piety had won the philosopher's hearty respect; next, with the same host, at Rijnsburg, near Leyden; then at the village of Voorburg, near The Hague; and finally at The Hague itself, where he died of tuberculosis in 1677 at the age of forty-four.

32 The bribe was allegedly 1,000 florins to be paid annually. The knife attack may have been an assassination attempt, though the record is unclear about the details—coming out of the synagogue or the theater, piercing Spinoza's skin or merely his clothing, and striking the face or the back of the neck.

Though Spinoza occasionally amused himself by sketching portraits, his chief relaxation was the grinding of lenses by way of earning the simplest of livelihoods. His solitude was a little broken sometimes by visits from eminent men, Leibniz, for example, and Henry Oldenburg, first secretary of the Royal Society of London, and was further relieved by a pretty extensive philosophic correspondence with a certain Amsterdam club of students who sought his views on various matters, and with numerous other persons, some of them men of great distinction.

What sustained Spinoza's spirit through all the years of his lonely toil was a triple power—a conviction, an aim, and a trust: (1) an adamantine conviction that "the greatest good is the knowledge of the union which the mind has with the whole of nature"; (2) an aim steadfastly directed at the attainment of that sovereign good; and (3) absolute trust in the power of Reason to reach the aim's realization. What he meant by that "conviction" none can know without reading the *Ethics* attentively.

Spinoza never held a teachership except for a time in Van den Ende's school when receiving Latin instruction from the proprietor. He was, however, when forty-one, offered the chair of philosophy at

"For the greatest good is for him to attain to the enjoyment of such a nature together with other individuals, if this can be. What is that nature I shall show in its proper place, namely, that it is the knowledge of the union which the mind has with the whole of nature."

SPINOZA,
TREATISE ON THE CORRECTION OF THE UNDERSTANDING

"From the third kind of knowledge necessarily arises the Intellectual Love of God. For, from this kind of knowledge arises perfect joy associated with the idea of God as its cause; that is Love of God, not as God is imagined to be present, but as God is understood to be eternal; and this is what I call the intellectual love of God."

SPINOZA,
ETHICS

Heidelberg. The tender, which was in highly complimentary terms, promised "the most perfect freedom in philosophizing, which the Prince [His Highness][33] feels assured you would not abuse by calling in question the established religion of the state." After deliberation Spinoza courteously declined the offer, basing his decision upon three grounds: (1) that he preferred "philosophic research" to "the teaching of young students"; (2) that, despite every precaution, his lectures would incur the risk of seeming to trespass beyond the prescribed limits of the promised freedom; and (3) that teaching in public would be incompatible with the tranquility which he loved and deemed essential to his meditations.

Of Spinoza's works, only two were published during his life, *The Principles of the Cartesian Philosophy*, and *A Treatise on Religion and the State*. This last, an "eloquently reasoned defense of liberty in speculative matters," at once produced a storm of opposition, was interdicted by the civil authorities, and soon thereafter placed on the Index Expurgatorius. In 1677, shortly after his death, appeared his *Opera Posthuma* containing the *Ethics*; two unfinished works, *On the Improve-*

33 Charles Louis, Elector Palatine.

ment of the Understanding, and a *Political Treatise*; selections from the author's correspondence; and a *Compendium on Hebrew Grammar.*

In accord with its Latin title (*Ethica Ordine Geometrico Demonstrata*), the *Ethics*, most famous of Spinoza's works, is fashioned on the pattern of Euclid's *Elements*, including the whole apparatus of axioms, postulates, definitions, theorems, demonstrations, corollaries, and scholia. Partly on this account and partly because of the exceeding tenuity, subtlety, and complexity of its subject-matter, Spinoza's *Ethics* is as hard to understand as Einstein's Doctrine of Relativity. That fact, though it may cause, cannot justify, neglect of it.

Was Spinoza a great mathematician? If the term is to designate only such as have made important contributions to so-called "pure" mathematics, the answer is No. But if it be applied also to those who have masterfully applied the mathematical method to no matter what kind of concrete subject-matter, the answer is Yes. For, in the domain of Ethics, the mathematical method became in Spinoza's hands a powerful instrument for both research and exposition. With the passions, desires, emotions, imaginations, wills, and concepts of men, Spinoza dealt as if with points, lines, planes, and their combinations. For him the Universe, or Nature, has a determinate *structure*, infinite and eternal, and thereby are all things determined with the same sort of necessity as that whereby the essential structure of a triangle determines its angle sum.

It is philosophy's aim and function thus to contemplate all things *under the aspect of eternity.* To achieve that is to achieve the highest blessedness, which is "the intellectual love of God."

GOTTFRIED WILEHLM LEIBNIZ

9

GOTTFRIED WILHELM LEIBNIZ

1646-1716

Gottfried Wilhelm Leibniz—probably the profoundest, most versatile, and most prolific of seminal thinkers in the annals of genius—was born at Leipzig, Germany, 1646. This was two years before the peace of Westphalia, ending the Thirty Years War [1618-1648], which had made the plight of Germany unspeakably desolate. Thus Leibniz's life began in what was (with the possible exception of the present) the darkest period in his country's history.[34] His father, a lawyer and professor of moral philosophy in the university, died when Wilhelm was six. In his mother, noted for her piety, her extraordinary patience in all trying circumstances, and her fostering care of tranquility and peace in all personal relations, some have thought to discern the source of that conciliatory spirit which in later years was to distinguish both the practical and the theoretic activity of her son in his strenuous endeavors to unite the great religious factions of his time and to conceive a true cosmology that should incidentally bring into harmonious relationship all the great philosophic insights of the ancient, medieval, and modern worlds.

Young Leibniz attended a Leipzig school but in the Leipzig of that time there was no school fit for such a boy. Yet he acquired a magnificent education because he had the fortune to have one magnificent teacher, a private one—Wilhelm himself. Latin he began at the age of eight; before he was twelve he wrote good Latin verses and had begun Greek. Before he was ten his father's

34 Millions died from the war and disease.

library, which had been kept under lock and key, was opened to him. There he lived in delightful communion with such as Plato, Aristotle, Cicero, Quintilian, Seneca, Pliny, Herodotus, Xenophon, the Roman historians, the Greek and Latin Fathers, the great scholastics and leading exponents of Protestant theology. Very noteworthy as a foretoken of his subsequent great *metaphysico* scientific dream of a Universal Mathematics—a dream recently rekindled in some able minds and now advancing toward a measure of realization in the ways of symbolic logic—is Wilhelm's attempt as a mere boy, under fifteen years, to reform the old logic (or logics) inherited from Aristotle, the Christian fathers and the medieval schools.

At the age of fifteen, he entered the University of Leipzig to study law, which for historical reasons had become in Germany a subject of deep interest. His legal studies did not keep him from here delving further into philosophy, ancient and scholastic; and so, though significant, it is not surprising that his bachelor's dissertation, submitted when he was barely seventeen, was a dexterous defense of nominalism, the doctrine that only individuals exist, classes, generalities, and universals being nothing but convenient constructs of the mind. At the same time by extra-curricular reading, he became aware of a great new birth of time—the advent of Natural Science—then increasingly manifest in the pioneering works of such as Copernicus, [Giordano] Bruno, Francis Bacon, [Tommaso] Campanella, [Gerolamo] Cardan[o], [William] Harvey, Kepler, Galileo, Descartes, Pascal, [Evangelista] Torricelli, and others. He thus discovered a fundamental defect in his education; for he soon saw that to understand this scientific movement demanded a degree of mathematical knowledge which, despite his inborn mathematical genius, now unknown even to himself, he had not acquired. In the near-by University of Jena, Erhard Weigel,[35] both mathematician and philosopher, was now lecturing on "natural law." Young Leibniz's intimate association with Weigel while attending his lectures in the summer of 1663 was a delightful and profitable experience, but no significant improvement in his too meager mathematical

35 Famous in this period for attempting to devise Pythagorean-type proofs for the existence of God, natural law, and the Holy Trinity.

equipment was to come till nine years later when more fortunate circumstances brought him face to face in Paris with genuine mathematicians and especially with Christiaan Huygens.

Meanwhile, he returned to Leipzig to concentrate on law. Owing to an astonishing pettiness, meanness, and jealousy in the faculty, the doctorate of law applied for by Leibniz in 1666, when he was twenty, was refused on the alleged ground that the applicant was too young. His dissertation proposing a new way—the historical way—of teaching law, was brilliant, and the degree refused at Leipzig was at once conferred by the University of Altdorf (in the free city of Nuremberg). At the same time young Leibniz was here offered a professorship of law, which he declined, saying that he had "very different things in view."

A man of profound learning in literature, history, philosophy, and law; author of several essays remarkable for their brilliance and penetration; a thinker driven and sustained by an ardent and boundless curiosity about all things, material or mental; a born genius of the first order: such was Leibniz at the age of twenty, and he had yet a half-century to live. So great was his activity during these years, so immense, manifold, and diversified his endeavors and achievements,

"Leibnitz, however anxious to throw out his mind upon the whole encyclopaedia of human research, yet did not forget to pay the price at which only any *right* to be thus discursive can be earned: he sacrificed to the austerer muses: knowing that God geometrizes eternally, he rightly supposed that in the universal temple Mathesis must furnish the master key which would open most shrines."

THOMAS DE QUINCEY,

1843

both theoretic and practical, that to set forth even a minimum of what one must know about him to see his figure in something like fair perspective among the great heroes of thought and action would require a long and elaborate essay. For such a survey would have to cover his work in logic, in mathematics, in mechanics and mechanical invention, in various branches of natural science, in theology, in philosophy (then embracing cosmology, metaphysics, ontology, epistemology, psychology, and esthetics), in jurisprudence, in statecraft and diplomacy, in history, in genealogy, in economics, in journalism, and in the planning, proposing, and founding of scientific academies. In all these fields he did original work, most of it important, much of it immortal. And—what seems well-nigh incredible—so mobile was his mind that at any time in his life he was able to work, and did work, effectively in several subjects or fields—as history, law, diplomacy, philosophy, mathematics, and others—simultaneously. No wonder that Thomas De Quincey called him "the last of the universals,"[36] not a very numerous class—Democritus, Aristotle, Leonardo da Vinci, and who besides?

Among Leibniz's achievements there are three which tower above all the rest like dominating peaks of a mountain range. Only a bare glance at these is here possible.

(1) One of the peaks is strictly mathematical. Noteworthy in this connection is a striking phenomenon observable at all stages in Leibniz's life—the shaping of his career by the interplay of interests often so divergent as to seem incompatible. Witness the following: Leibniz's essay advocating the historical method in the study and teaching of law wins the admiration of the statesman [Johann Christan von] Boyneburg; upon his advice the essay is brought to the attention of the Elector of Mainz; on invitation of this high official, bent on maintaining the security of the German Empire, sorely

36 Not De Quincey and not this exact phrase. William Gladstone: "Let us bow our heads to the inevitable; the day of encyclopædic learning has gone by. It may perhaps be said that that sun set with Leibniz." Alternative Gladstone: "As we now stand, no individual man holds or can hold that relation to universal knowledge, which was held by Dante, or by Bacon, or by Leibniz."

disturbed by dissensions within and by threats of aggression from without, Leibniz enters the Elector's service, first as reviser of the statute-book and then as general counselor in politics, statecraft, and diplomacy; in 1670 writes, in German, his memorable *Thoughts on Public Security*; is consequently sent in 1672 on a diplomatic mission to Paris, later to London; in these cities meets philosophers, scientists, and mathematicians; is mathematically aroused by the latter, especially by Huygens; plunges into mathematical research, at first under the direction of Huygens, and soon thereafter, concurrently with Newton and independently of him, invents the Infinitesimal Calculus, "the most powerful instrument of thought yet devised by the wit of man."

(2) A second of the peaks is also mathematical if the adjective be taken broadly. It is the dream, already mentioned, of a Universal Mathematics—*Ars Combinatorial* as Leibniz called it—and, in essence, is this: that there are simple ideas and simple truths, that these are discoverable and constitute a manageably small set of elements, and that from them all other ideas and truths, however composite, are derivable by combination in accord with simple rules and by help of a symbolic

"The soul follows
its own laws,
and the body likewise
follows its own laws;
and they agree
with each other in
virtue of the
pre-established
harmony between
all substances,
since they are all
representations
of one and the same
universe."

LEIBNIZ,
THE MONADOLOGY

language—*characteristica universalis*—not too hard to invent. This unsurpassed dream, then pooh-poohed by many, neglected by all, never abandoned by its author nor fully elaborated, figuring only sketchily in many letters and articles, published and unpublished, is today a driving force in the heart of modern logic. Is it completely realizable? Such a dream is a genuine ideal. Being such, it is not a goal—a thing to be reached—but a flying perfection to be endlessly pursued.

(3) Finally, there is the monadology, a singularly ingenious and noble world theory based on the assumption that the cosmos—all reality—is composed of an infinitude of individuals (monads), extensionless, not material but psychic, all living, mutually independent, constituting a hierarchy of spiritual dignities, each representing all, all striving in preestablished harmony to attain to the perfection of the Supreme Monad, or God.

Leibniz wrote no single work setting forth his philosophy as a whole. It exists only in precious fragments found in many published and unpublished papers and in his voluminous correspondence.

At the age of seventy, the greatest of seminal thinkers—immensely famous in the prime of his life, seemingly forgotten and certainly neglected in his closing years—died at Hanover, the scene of his labors for forty years. A single mourner, his secretary, followed his body to the grave.

GEORGE BERKELEY

10

GEORGE BERKELEY

1684-1753

George Berkeley—famous as the reputed founder-in-chief of modern philosophical Idealism—was born March 12, 1684, in a cottage at Kilrin near Thomastown, Ireland. His father, William Berkeley, was an Englishman. The son sometimes called himself an Irishman but never a "native Irishman." It is said on good authority that [Jonathan] Swift introduced George to Lord Berkeley of Stratton in the following words: "Here is a fine young gentleman of your family. I can assure your Lordship it is a much greater honor to *you* to be related to him than it is to him to be related to you."

Of George's boyhood only a very little is known but that little is significant. From twelve to sixteen he was a pupil in Kilkenny School, where he was noted for precocity. In his *Commonplace Book*, exceedingly rich in autobiographical material but not published till 1871, Berkeley says: "I was distrustful at the age of eight and consequently by nature disposed for new doctrines." But his inborn critical spirit, manifest so early, was directed against himself quite as much as against others. "I am young," says he in the youthful diary already alluded to:

> I am an upstart, I am a pretender, I am vain...But one thing I know I am not guilty of. I do not pin my faith on the sleeve of any great man. I act not out of prejudice or prepossession. I do not adhere to any opinion because it is an old one, a reviv'd one, a fashionable one, or one that I have spent much time in the study and cultivation of.

In his youth as throughout his life Berkeley had an insatiable craving for facts, for knowing things as they really

"In this respect I would
with pleasure do justice to
the Memory of a very great
tho' singular sort of a Man,
Dr. Berkeley, better known
as a Philosopher, and
intended Founder of an
University in the Bermudas,
or Summer-Islands,
than as Bishop of Cloyne
in Ireland. An Inclination
to carry me out on that
Expedition, as one of
the young Professors on
his new Foundation,
having brought us often
together, I scarce
remember to have
conversed with him on
that Art, liberal or
mechanic, of which he
knew not more than
the ordinary Practitioners.
With the widest Views,
he descended into
a minute Detail, and
begrudged neither
Pains nor Expence for
the means of Information.
He travelled thro' a
great part of Sicily on
foot; clambered over the
Mountains and crept
into the Caverns to
investigate into its natural
History, and discover
the Causes of its
Volcanoes: and I have
known him sit for
Hours in Forges and
Founderies to inspect their
successive Operations."

THOMAS BLACKWELL,

1755

are. When fourteen he explored the cave of Dunmore,[37] and wrote an admirable report of his findings. "I have known him," said a friend, "to sit for hours in Forges and Foundries to inspect their successive operations."

In 1700, Berkeley entered Trinity College, Dublin, at the age of sixteen. For a young student with open mind, candid, keen, eager, enquiring, independent, the time was propitious. Scholasticism had had its long day and was receding. The intellectual atmosphere, both within and without academic circles, was astir with a new philosophy, a new kind of thinking. Young Berkeley was quick to feel the sting of the "new ideas," the ideas of such as Descartes and [Nicolas] Malebranche,[38] Locke and Newton, Huygens and Galileo, Copernicus and Leibniz, Spinoza and Kepler, and he soon became a student of mark—some thought him a great genius, others deemed him a fool. In four years he gained the bachelor's degree, in three more the master's, and at the same time won a fellowship by an examination "passed with great distinction." He held a tutorship in the College from 1707 to 1724, served as sub-lecturer in 1710, as junior dean in 1710 and 1711, and in 1712 was made junior lecturer in Greek.

Berkeley's first intellectual love was mathematics, and when twenty-one he wrote in Latin two mathematical tracts, which were published two years later anonymously. These juvenile essays are noteworthy on several accounts: (1) for their high praise of "algebra" as "the highest pinnacle of human knowledge, [the] kernel and key of all mathematics";[39] (2) for their warm appreciation of the esthetic pleasure attending mathematical manipulations; (3) for their admiration of "the clear light of demonstration"; (4) for their signalizing of the moral value of mathematical study in "abstract[ing] the mind from the senses, sharpen[ing] and confirm[ing] the talents"; and (5) especially for their insistence upon the capital importance of research in the logical foundations of mathematics.

In the item last mentioned one may discern a kind of harbinger of the bold and subtle criticisms which Berkeley, twenty-seven years later, launched in

37 Located in Ireland, site of a 10th-century Viking assault.

38 Controversial French priest, who blended elements of St. Augustine and Spinoza.

39 *"supremus cognitionis humanæ apex, universae matheseos nucleus et clavis."*

his famous *Analyst* first against certain ideas and processes employed by [the Marquis de] L'Hôpital[40] and other continental mathematicians in their expositions of the Calculus and then against Newton himself. Those criticisms quickly produced a veritable furor and are still sometimes pooh-poohed as rash outgivings of a presumptuous ignoramus. Yet some of the criticisms are just, being solidly based on a perfectly sound lemma admirably formulated by Berkeley thus.

> If with a view to demonstrate any proposition, a certain point is supposed, by virtue of which certain other points are attained; and such supposed point be itself afterwards destroyed or rejected by a contrary supposition; in that case, all the other points attained thereby, and consequent thereupon, must also be destroyed and rejected, so as from then forward to be no more supposed or applied in the demonstration.

Berkeley's early enthusiasm for mathematics led him to believe for a time that ethics, the existence of God, the immortality of the soul, and other kindred matters could be established mathematically. But when he subsequently discovered that mathematics could be logically rigorous only because its propositions were abstract, and meaningless in the sense that their symbols were quite devoid of concrete content, his enthusiasm for the subject, though it never quite perished, suffered much abatement. And his initial admiration for scientific method as a means to knowledge of concrete reality underwent a like change.

In 1712-1713, Berkeley spent some months in London where his abilities, scholarship, magnetism, and charm won him the friendship of such as Swift, Addison, Steele, Pope, Gay, and Arbuthnot,[41] and made him *persona grata* at court. During nine months of 1713-1714 he traveled on the continent as chaplain to Lord Peterborough, and again in 1715-1720 as tutor to the son of Bishop Ashe. In 1721 he returned to Ireland as

40 Paid mathematician Johann Bernoulli for his calculus-related discoveries.

41 Jonathan Swift (*Gulliver's Travels*), Joseph Addison and Richard Steele (co-founders of *The Spectator*), Alexander Pope ("to Err is Human; to forgive, Divine"), John Gay (English poet and dramatist), and John Arbuthnot (Scottish polymath).

"VERSES ON AMERICA"

The Muse, disgusted at an age and clime
  Barren of every glorious theme,
In distant lands now awaits a better time,
  Producing subjects worthy fame:

In happy climes, where from the genial sun
  And virgin earth such scenes ensue,
The force of art by nature seems outdone,
  And fancied beauties by the true:

In happy climes, the seat of innocence,
  Where nature guides and virtue rules,
Where men shall not impose for truth and sense
  The pedantry of courts and schools:

There shall be sung another golden age,
The rise of empire and of arts,
The good and great inspiring epic rage,
The wisest heads and noblest hearts.
Not such as Europe breeds in her decay;
Such as she bred when fresh and young,
When heavenly flame did animate her clay,
By future poets shall be sung.

Westward the course of empire takes its way;
The first four Acts already past,
A fifth shall close the Drama with the day;
Time's noblest offspring is the last.

GEORGE BERKELEY,
MISCELLANY

chaplain to the Duke of Grafton and was appointed divinity lecturer and university preacher. In 1722 he became dean of Dromore, without duties, and served the university as senior proctor and teacher of Hebrew. Meanwhile his fame was enhanced by a number of publications, among them his *Dialogues* and fourteen contributions to Steele's *Guardian*.[42]

In 1724, Berkeley sought and obtained the Deanship of Derry, believing that the distinction of this title and the emoluments of that richest of Irish deaneries would help him to realize a great dream that had taken control of his head and heart—the establishment in the salubrious climate of Bermuda of a College which, being remote from the established vices of the Old World, would, he hoped, increasingly become a center of virtue and enlightenment, and eventually create a genuine civilization in "the land of the setting sun." Four years of strenuous exertion won for his project the cooperation of many influential men and women, a charter, and the government's promise of twenty thousand pounds. There upon, without waiting for payment, Berkeley and his wife, with some friends, set sail in September, 1728. Desiring to arrange for providing the projected college with fresh meat, they stopped at Rhode Island. The government failed to keep its promise, and Berkeley, after devoting three years to quiet study in Rhode Island, abandoned his Bermuda dream and returned to England.

In 1734, Berkeley became Bishop of Cloyne. After a service there of eighteen years he moved with his family to Oxford where he died suddenly and peacefully January 14, 1753, at the age of sixty-nine.

Berkeley was too self-critical to build a philosophical system. In general, the later works involve revision, sometimes abandonment, of positions maintained in earlier ones. They thus show growth, development, progress. Hence it is best to read them in the order of their appearance. Berkeley was aware of this as he pointed out in a letter [two letters] of 1730 to his friend and pupil, Samuel Johnson[43] (later the first president of King's College, now Columbia University):

> I could wish that all the things I have published on these philosophical subjects were read

42 Closed down in 1713 after less than a year. Unrelated to the present-day *Guardian*.

43 Samuel Johnson, unrelated to the English essayist "Dr Johnson" of the same name.

> in the order wherein I have published them, once, to take in the design and connexion of them, and a second time with a critical eye, adding your own thought and observation upon every part as you went along...
>
> What you have seen of mine was published when I was very young and without doubt hath many defects...I do not therefore pretend that my books can teach truth. All I hope for is that they may be an occasion to inquisitive men in discovering truth, by consulting their own minds, and looking into their own thoughts.

Whether justly or unjustly Berkeley's fame and influence have been mainly due to what philosophers have designated Berkeleyan Idealism. The best initial clue to this famous doctrine is found in Berkeley's own words:

> Some truths there are, so near and obvious to the mind, that a man need only open his eyes to see them. Such I take this important one to be, viz., that all the choir of heaven and furniture of the earth, in a word, all those bodies which compose the mighty frame of the world, have not any subsistence without mind; that their *being* is to be perceived or known; that consequently so long as they are not actually perceived by me, or do not exist in my mind, or that of any other created spirit, they must either have no existence at all, or else subsist in the mind of some Eternal Spirit...
>
> When I deny sensible things an existence out of the mind, I do not mean my mind in particular, but all minds. Now, it is plain they have an existence exterior to my mind; since I find them by experience to be independent of it. There is therefore some other mind wherein they exist, during the intervals between the time of my perceiving them: as likewise they did before my birth, and would do after my supposed annihilation. And, as the same is true with regard to all other finite created spirits, it necessarily follows there is an

omnipresent eternal mind, which knows and comprehends all things, and exhibits them to our view in such a manner, and according to such rules, as He Himself hath ordained, and are by us termed the *laws of nature*.

IMMANUEL KANT

11

IMMANUEL KANT

1724-1804

IMMANUEL KANT—CREATOR AND KING of the Critical Philosophy (so called) of the nineteenth century—was born, the second of six children, April 22, 1724, in the Prussian city of Königsberg, which then contained about fifty thousand inhabitants and where, save for a single journey of forty miles, he spent all the days of his life of eighty years. His paternal grandfather was a Scotchman, who had emigrated to Germany about a century before Immanuel was born. The father, who was a saddler, and his mother, "a woman of exalted character, and of intellectual accomplishments beyond her rank," were ardent Pietists,[44] and hoped that Immanuel might become a truly religious force as a Lutheran clergyman. When the boy was thirteen his mother died; but she had deeply impressed him for, though Kant as an adult attended no church, "[he] never spoke of his mother to the end of his life without the utmost tenderness and acknowledgment of his great obligations to her maternal care."

With some assistance from a near relative the parents were enabled, despite their slender means, to start Immanuel on the way to a liberal education, without, however, having any inkling as to what he would eventually become. When a child they sent him to a charity school, and later, when he was about ten, to the Royal Academy to prepare him for theology. Here he gained distinction, and a permanent interest, in Greek and Latin classics, especially in Lucretius. At sixteen Kant became a student in the University of Königsberg where

---

44 Pietism was a Protestant reform movement emphasizing personal and "practical" Christianity ("*ecclesiola in ecclesia*," churches in the church) rather than theological orthodoxy.

he manifested a strong predilection, not for theology, but for mathematics and physics, and, when twenty-two, printed his first paper, dealing acutely with the question, long debated by Cartesians and Leibnizians, as to the proper measure of a "living force."[45] His father having died at this time, Kant's university studies were much disturbed by his having to earn his own living for nine years by means of private tutoring, an occupation which he heartily disliked.

At length he was enabled by a friend's generosity to resume his university career, and in 1755 he gained the doctorate and qualified as *privatdozent*. In this lowly academic position Kant had to remain for fifteen years though his fame as a writer and lecturer had been steadily increasing. Professorial appointments at other universities he declined.

In the course of his long service in the relatively humble post of *privatdozent*, Kant dealt so ably, both in his lectures and in his publications, with so great a variety of important questions, both scientific and philosophic, that to any discerning "official authority" it should have been clearly evident that here was a thinker of epoch-making power and that to deny him due recognition was an unpardonable academic sin. Yet, incredible as it may seem, Kant had reached the age of forty-six before such recognition came, when in 1770 he was appointed to the chair of mathematics and shortly thereafter to that of logic and metaphysics. This was years after he had, when twenty-two, virtually settled the above-mentioned Cartesian-Leibnizian dispute respecting the measure of *vis viva*; after he had, when thirty, more than a century before [Charles-Eugène] Delaunay, detected the retarding effect of moon-caused tides on the rotation of the earth; after he had, when thirty one, formulated the hypothesis which subsequently became famous as the Nebular

45 The debate over *vis viva* is obscure today. At the time, the definitions and relationships between concepts such as force, mass, and velocity were all unclear and contentious. To oversimply, Descartes held that the "quantity of motion" is proportional to mass multiplied by velocity, while Leibniz held that the "living force" is proportional to mass multiplied by velocity squared. The former evolved into the modern notion of "momentum" and the latter into "kinetic energy" (with minor modification).

Hypothesis of [Pierre-Simon] Laplace;[46] after he had made, when thirty-two, important contributions to the theory of earthquakes and the now received theory of winds;[47] and not only years after such weighty contributions to *science* but also after the publication of no less than half a score of very suggestive and penetrating memoirs in metaphysics and other branches of *philosophy*.

But the foregoing works, though they witness to both the scientific and the philosophic genius of their author, were only preliminary and preparatory. The works that made Kant's name one of the very greatest in the history of thought were produced, not before but after, his long-delayed elevation to the Königsberg professorship. Of these the first one to appear was the *Critique of Pure Reason*, commonly thought to be the greatest and undoubtedly the most famous of all Kant's writings. Its date is 1781. It cost the author eleven years of the most arduous and methodical meditation following his professorial appointment plus the

"It was not only in the character of a companion that Kant shone, but also as a most courteous and liberal host, who had no greater pleasure than in seeing his guests happy and jovial, and rising with exhilarated spirits from the mixed pleasures–intellectual and liberally sensual–of his Platonic banquets."

THOMAS DE QUINCEY,
1827

46 French astronomer who argued that our solar system formed from a massive gas cloud.

47 The Lisbon earthquake of 1755 prompted Kant to write several essays in 1756. The same earthquake influenced Voltaire's views on the problem of evil: it occurred on All Saints' Day.

more or less conscious preparation of the preceding forty-six years. Kant was now fifty-seven. Of all philosophers he seems to have been the most deliberate. In 1787 appeared a second and somewhat revised edition of this work.

In about a decade after its appearance the *Critique* was being expounded in all the leading universities of Germany; had been anxiously examined and some times approved by distinguished theologians, Protestant and Catholic; drew students to Königsberg as to a holy shrine; swiftly invaded other lands; and thereafter ruled the serious philosophizing of the Western world for a hundred years. Today its influence is still felt everywhere.

What is Kant's doctrine? Here it is impossible to analyze it or even to summarize it briefly. It would not be difficult, were it advisable, to set down a dozen or a score or more of sentences each of them purporting to be or to represent some cardinal thesis among Kant's teachings. (One such statement, for example, would be that, according to Kant, Space and Time are forms, but not objects, of perception; another would be that freedom of will, immortality, and the existence of God cannot be established by the Pure Reason but can be by the Practical Reason; a third might be that a natural science is a science only insofar as it is mathematical.) But the sentences could be true statements of Kantian doctrine only if they were rightly interpreted. That *if* is, however, very embarrassing. For the sentences could not be rightly interpreted without the light which nothing but a careful examination of Kant's writings can yield.

But Kant's works are, proverbially, hard to read understandingly. Herbert Spencer's candid confession[48] that he could not understand them has doubtless consoled many a would-be philosopher. Nevertheless, despite its repellent jargon of technical terms and its ponderous style, only seldom relieved by a too brief passage of genuine beauty and light, even the *Critique*, hardest of all Kant's works, can be understood. Certainly it is not

48 Spencer: "My knowledge of Kant's writings is extremely limited...I read the first few pages enunciating his doctrine of Time and Space; my peremptory rejection of which caused me to lay the book down. Twice since then the same thing has happened; for, being an impatient reader, when I disagree with the cardinal propositions of a work I can go no further."

harder to understand than the *Ethics* of Spinoza. It was mastered, for example, by Charles Sanders Peirce but only by his devoting "two hours a day to it for more than three years, until he almost knew the whole book by heart, and had critically examined every section of it." Later, Peirce called Kant "the king of modern thought."

For English-speaking students, the best approach to Kant is through English translations and more or less popular English expositions.

Kant's physical stature was small, just slightly above five feet; his breast was flat, almost concave; his appearance frail. But his senses were quick, delicate, and accurate; and his health was nearly perfect owing to strictness of regimen and an almost incredible methodicalness of living—his neighbors could set their clocks by his punctual comings forth for his daily walk.

He was an omnivorous reader, not so much in the history of philosophy as in general history, science, books of travel, *belles lettres* of the more didactic type, and in newspapers and political works. His appreciation was, however, not universal; he seemed somewhat insensitive to the beauties of nature, never sailing on the Baltic nor exploring the environs of his native city. Had he had a love for music his writings might have had some of the limpidity and charm of [Arthur] Schopenhauer.[49] Curiously enough his lectures were, unlike his written discourse, distinctly charming. Based on a few notes, they were extemporaneous, enlivened by anecdotes, and, though his voice was feeble, rose sometimes to the height of genuine eloquence. "One of his practical principles was to attend most to those pupils who were of middle ability; the dunces, he said, were beyond all help, and the geniuses would help themselves."

Kant never married. He has been reproached for holding too low an ideal for women. But he always revered the memory of his mother, and in his last illness was nursed with great tenderness by his sister.

One of the most memorable of Kant's ethical utterances has been rendered by Lord Houghton[50] in the following couplet:

---

49 Intellectual heir to Kant.

50 Richard Monckton Milnes (1809-1885), British poet-politician.

Two things I contemplate with
ceaseless awe:
The stars of Heaven, and Man's
sense of law.

The "Law" here meant is the law of Duty, an inner law—the categorical imperative: "Act always on such a maxim as thou canst at the same time will to be a universal law." It is very noteworthy that the great philosopher, though "king of modern thought," yet held himself and all humans to be subject to an authority transcending the domain of Pure Reason.

CHARLES SANDERS PEIRCE

12

CHARLES SANDERS PEIRCE

1839-1914

Charles Sanders Peirce—regarded by competent critics as the most original as well as the profoundest and most significant of American philosophical thinkers—was born September 10, 1839, at Cambridge, Massachusetts. His father, Benjamin Peirce, a pioneer in American mathematical research, was long a distinguished professor at Harvard, first of mathematics and natural philosophy and then of mathematics and astronomy. Subsequently, Benjamin's eldest son, James Mills Peirce, was a professor of mathematics at Harvard for more than two score years.

Though Charles was graduated at Harvard and later at the Lawrence Scientific School, he said of his father: "He educated me, and, if I do anything, it will be his work."

For some years Peirce, who throughout his life was keenly interested in science, held a post in the Coast and Geodetic Survey[51] where he gained some distinction in geodetic research.

Like his father and brother, Charles Peirce, too, was a mathematician, but his mathematical work was mainly done in the domain of Logic. Logic was his chief concern.

Logic and mathematics, as Peirce conceived them, are essentially different things. He adopted his father's famous definition of mathematics as "the science which draws necessary conclusions" but he pointed out that to draw necessary conclusions is one thing; that to draw conclusions is a second and vaster thing, for most conclusions are not necessary;

51 Merged with the National Oceanic and Atmospheric Administration in 1970. "Geodesy" means measuring Earth.

"In order to be deep it is requisite to be dull."

CHARLES SANDERS PEIRCE, 1903

and that to investigate the conditions for drawing conclusions (whether necessary or only "probable") is a third and still vaster thing. This last, said Peirce, is Logic. Logic is the *science of drawing conclusions.*

In view of that conception, it is easy to understand why Peirce regarded Logic as inferior to no other subject in difficulty or in dignity or in scope or in human significance. To research in the field of Logic he brought a quite unsurpassed measure of the requisite type of genius, and devoted to it greater energy perhaps and more years than any other man in the annals of science. His logical researches inevitably led him to deal with almost every fundamental kind of question met with in the history of thought.

In 1880 [1879], at Johns Hopkins,[52] Peirce lectured on Logic under the auspices of the department of Mathematics, of which James Joseph Sylvester was then the famous head. One day Sylvester said to a beloved mathematical student of his: "You have attended Mr. Peirce's lectures. Tell me your impression of them." At some length the student responded that they were always substantial, never trite, often very subtle,

52 Johns Hopkins was founded only in 1876.

frequently so profound or obscure that listeners were unable to understand, "but," added the student, "Mr. Peirce is certainly a genius." Thereupon Sylvester said impulsively: "Well! If he is a genius, isn't that enough? Isn't it geniuses that we want here?"[53]

Aware of his originality and superb scholarly equipment some have wondered why Peirce was not appointed to a professorship at Johns Hopkins or at Harvard where his family name had attained high prestige or elsewhere. Was he perhaps too original or too independent or too eccentric or too unconventional to be academically clubable? "I am a man," he once wrote, "of whom critics have never found anything good to say." The critics are now dead, most of them, and mostly forgotten, while Peirce's fame is growing.

Perhaps his country's academic neglect of him was a blessing in disguise. Doubtless the solitude to which he was thus constrained was favorable to meditation. But there were disadvantages. "[A]ll my life," he wrote, "my studies have been cruelly hampered by my inability to procure necessary books." At all events most of his major work was done in the privacy of his home in Milford, Pennsylvania, where he lived many years, often without the means of comfortable subsistence, and where he died in 1914 at the age of seventy-five. There his widow remained until her death twenty years later. Joseph Jastrow has written a poignantly beautiful tribute to Mrs. Peirce's devotion and loyalty to the memory of the great man who was her husband.

Peirce has been rightly called a *seminal* thinker—a thinker of seed thoughts. Out of such a thought of his sprang the philosophy called Pragmatism. He planted the seed as a maxim of Logic. Here it is, twice expressed in his own words:

> Consider what effects, that
> might conceivably have practical
> bearings, we conceive the object
> of our conception to have. Then,
> our conception of these effects
> is the whole of our conception
> of the object.

---

53 Likely an anecdote told directly to C.J.K. by that "beloved mathematical student," who remains nameless. It could have been Fabien Franklin, Christine Ladd, or Joseph Jastrow, all of whom wrote obituaries upon Pierce's passing. Incidentally, Franklin and Ladd married.

> In order to ascertain the meaning of an intellectual conception one should consider what practical consequences might conceivably result by necessity from the truth of that conception; and the sum of these consequences will constitute the entire meaning of the conception.

Planted in 1878, the seed lay dormant for twenty years. Then suddenly it burst forth a stirring philosophy elaborated and sponsored by William James and others. These Peirce thought a bit too eloquent and "lively" to represent his idea. "[T]o be deep," said he, "it is requisite to be dull." So he decided to rename his idea. Here is his delicious announcement:

> So then, the writer, finding his bantling "pragmatism" so promoted feels that it is time to kiss his child good-by and relinquish it to its higher destiny; while to serve the precise purpose of expressing the original definition, he begs to announce the birth of the word "pragmaticism," which is ugly enough to be safe from kidnappers.

Peirce was a pioneer in the study of manifolds or classes, finite and infinite. A digest of what he wrote on this subject would make an excellent master's essay in philosophy or mathematics. His way of discriminating the terms *finite* and *infinite* is original and characteristic. Being primarily a logician and consequently interested in argument forms, he defined a class to be finite or infinite according as, when applied to it, the "syllogism of transposed quantity" is or is not valid. An example of such a syllogism is:

> Every Hottentot kills a Hottentot;
> No Hottentot is killed by more than one Hottentot;
> Therefore every Hottentot must be killed by a Hottentot.[54]

The term propositional function was given to philosophy by Bertrand Russell but the important idea denoted by it was

54 "Hottentot" is a colonial-era Dutch term, now derisive, referring to tribes in southern Africa.

familiar to Peirce, who called it a *rheme*, before Russell was born.

The like may be said of those curious phenomena called paradoxes, many of which are listed and discussed by [Alfred North] Whitehead and [Bertrand] Russell in their *Principia Mathematica.* Long before the appearance of that work numerous such paradoxes were noted and acutely discussed by Peirce under the caption *Insolubilia.*

Another field in which Peirce did extensive pioneering work is that of the logic of Relations, which he called the logic of Relatives. As to its importance, for a logician to neglect Relations, as was done by Aristotle and all other logicians previous to [Augustus] De Morgan, said Peirce, as if a geographer pretending to describe the earth were to confine his attention to Nantucket.

If the seed thoughts to be found in Peirce's writings were gathered together, they would truly constitute an astonishingly large and dazzling collection of gems. In this short essay there is room for mention of only one more of them, one whose significance may not be duly appreciated for a century or a thousand years. It is called *Tychism.*

Tychism is a concept or a doctrine of *evolution.* It is not the usual conception of the development of the organic and inor-

"For this subject, although always recognised as an integral part of logic, has been left untouched on account of its intricacy. It is as though a geographer, finding the whole United States, its topography, its population, its industries, etc., too vast for convenient treatment, were to content himself with a description of Nantucket. This comparison hardly, if at all, exaggerates the inadequacy of a theory of reasoning that takes no account of relative terms."

CHARLES SANDERS PEIRCE

ganic worlds under an assumed beginningless and endless and universal reign of Law. It is the conception of a beginningless and *endless evolution of Law itself*: a doubly infinite process running, so to say, from virtual Chaos—a backward limit—towards perfect Cosmos—a forward limit that time, however prolonged, can never reach.

To gain a just sense of the intellectual stature of Charles Sanders Peirce one must examine his writings attentively. That this man, who immeasurably increased the intellectual wealth of the world, was nevertheless almost permitted to starve in what was in his time the richest and vainest of lands is enough to make the blood of any decent American boil with chagrin, indignation, and vicarious shame.

CONCLUSION

At length, when thought, in its progress along any one of the many courses, has reached a high degree of refinement, precision and certitude, then and thereafter, but not before, we call it mathematical thought; it has undergone a long process of refining evolution and acquired at length the name of mathematics; it is not, however, the creature of its name; what is called mathematics has been long upon the way, owning at previous stages other designations—common sense, practical art perhaps, speculation, theology it may be, philosophy, natural science, or it may be for many a millennium no name at all. Is it, then, only a question of names? In a sense, yes: the ideal of thought is rigor; mathematics is the name that usage employs to designate, not attainment of the ideal, for it cannot be attained, but its devoted pursuit and close approximation. But this is not the essence of the matter. The essence is that all thought, thought in all its stages, however rude, however refined, however named, owns the unity of being human spiritual activities are one. Mathematics thus belongs to the great family of spiritual enterprises of man. These enterprises, all the members of the great family, however diverse in form, in modes of life, in methods of toil, in their progress along the way that leads towards logical rectitude, are alike children of one great passion. In genesis, in spirit and aspiration, in motive and aim, natural science, theology, philosophy, jurisprudence, religion and art are one with mathematics: they are all of them sprung from the human spirit's craving for invariant reality in a world of tragic change; they all of them aim at rescuing man from "the blind hurry of the universe from vanity to vanity": they seek cosmic stability—a world of abiding worth, where the broken promises of hope shall be healed and infinite aspiration shall cease to be mocked.

Such has been the universal and dominant aim and such are the cardinal forms that time has given its prosecution.

"To take into
the inmost shrine
of the soul the
irresistible forces
whose puppets we
seem to be—Death
and change,
the irrevocableness
of the past, and the
powerlessness of Man
before the blind
hurry of the universe
from vanity to
vanity—to feel these
things and know
them is to
conquer them."

BERTRAND RUSSELL,

1903

And now we must ask: What have been the fruits of the endless toil? What has the high emprize won? And what especially, have been the contributions of mathematics to the total gain? To recount the story of the spirit's quest for ageless forms of reality would be to tell afresh, from a new point of view, the history of human thought, so many and so diverse are the modes or aspects of being that men have found or fancied to be eternal. Edifying indeed would be the tale, but it is long, and the hour contracts. Even a meager delineation is hardly possible here. Yet we must not fail to glance at the endless array and to call, at least in part, the roll of major things. But where begin? Shall it be in theology? How memory responds to the magic word. "The past rises before us [me] like a dream." As the long succession of the theological centuries passes by, what a marvelous pageant do they present of human ideals, contrivings and dreams, both rational and superrational. Alpha and Omega, the beginning and ending, which is, which was and which is to come [Revelation 1:8]; I Am That I Am [Exodus 3:14]; Father of lights with which is no variableness, neither shadow of turning [James 1:7]; the *bonitas, unitas, infinitas, immutabilitas* of Deity; the undying principle of soul; the sublime hierarchy of immortal angels, terrific and precious, discoursed of by sages, commemorated by art, feared and loved by millions of men and women and children: these things may suffice to remind us of the invariant forms of reality found or invented by theology in her age-long toil and passion to conquer the mutations of time by means of things eternal.

But theology's record is only an immense chapter of the vastly more inclusive annals of world-wide philosophic speculation running through the ages. If we turn to philosophy understood in the larger sense, if we ask what answers she has made in the long course of time to the question of what is eternal, so diverse and manifold are the voices heard across the centuries, from the East and from the West, that the combined response must needs seem to an unaccustomed ear like an infinite babel of tongues [Genesis 11:9]: the Confucian Way of Heaven; the mystic Tao, so much resembling fate, of Lao Tzu and Chuang Tzu; Buddhism's inexorable spiritual law of cause and effect and its everlasting extinction of individuality in Nirvana—the final blowing out of consciousness and character alike; Ahura Mazda, the holy One, of Zarathustra [Zoroaster]; Fate, especially in the Greek tragedies and Greek religion, the chain of causes in nature, "the compulsion in the way things grow," a fine thread running

"Does Heaven speak? The four seasons pursue their courses, and all things are continually being produced, but does Heaven say anything?"

CONFUCIUS,
THE ANALECTS

"Misery!–happiness is to be found by its side! Happiness!–misery lurks beneath it! Who knows what either will come to in the end?"

LAO TZU,
TAO TE CHING

"One path only is left for us to speak of, namely, that *It is*. In it are very many tokens that what is, is uncreated and indestructible, alone, complete, immovable and without end . . . Wherefore it is not permitted to what is to be infinite; for it is in need of nothing; while, if it were infinite, it would stand in need of everything."

PARMENIDES,
5TH CENTURY B.C.

through the whole of existence and binding even the gods; the cosmic matter, or το ἄπειϱον [the apeiron] of Anaximander; the cosmic order, the rhythm of events, the logos or reason or nous, of Heracleitus; the finite, space-filling sphere, or One, of the deep Parmenides; the four material and two psychic, six eternal, elements, of Empedocles; the infinitude of everlasting mind-moved simple substances of Anaxagoras; the infinite multitude and endless variety of invariant "seeds of things" [*semina rerum*] of Leucippus, Democritus, Epicurus and Lucretius, together with their doctrines of absolute void and the conservations of mass and motion and infinite room or space; Plato's eternal world of pure ideas; the great Cosmic Year[55] of a thousand thinkers, rolling in vast endlessly repeated cycles on the beginningless, endless course of time from eternity to eternity; the changeless thought-forms of Zeno, Gorgias and Aristotle; Leibnitz's indestructible, pre-established harmony;

55 C.J.K.: "[T]he philosophy of the cosmic cycle regards all the changes of which the universe is capable as constituting an immense but finite and closed system of transformations, which follow each other in definite succession, like the spokes of a gigantic revolving wheel, until all possible changes have occurred in the lapse of a long but finite period of time—called a cosmic cycle or a cosmic year—whereupon everything is repeated precisely, and so on and on without end."

Spinoza's infinite unalterable substance; the Absolute of the Hegelian school; and so on and on far beyond the limits of practicable enumeration. This somewhat random partial list of things will serve to recall and to represent the enormous motley crowd of answers that the ages of philosophic speculation have made to the supreme inquiry of the human spirit: what is there that survives the mutations of time, abiding unchanged despite the whirling flux of life and the world?

"When life comes, it cannot be declined; when it goes, it cannot be detained."

CHUANG TZU,
ZHUANGZI

Thou deep Base of the World, and thou high Throne
Above the World, whoe'er thou art, unknown
And hard of surmise, Chain of Things that be,
Or Reason of our Reason; God, to thee
I lift my praise, seeing the silent road
That bringeth justice ere the end be trod
To all that breathes and dies.

EURIPIDES,

5TH CENTURY B.C.

DISCUSSION QUESTIONS

1. Keyser notes that, according to Plato, Pythagoras was beloved to his disciples because he taught them "a certain way of life." Is there still a form of life appropriate to pursuing science and mathematics? Is there a kind of discipline needed?

2. In Plato's *Republic*, the philosopher-kings receive a mathematical education as a preparation for philosophy. The mathematical education consists of arithmetic (study of number in itself), geometry (study of number in space), music (study of number in time), and astronomy (study of number in space and time). This four-part mathematical curriculum appears to be one inspiration for the *quadrivium* within the seven liberal arts in the Middle Ages. What role does mathematics play in a liberal or liberating education? Is Plato correct to place mathematics as preparation for and therefore subordinate to philosophy? Does Keyser agree?

3. Keyser mentions Roger Bacon as the inventor of gunpowder. In his *Novum organon*, Francis Bacon names three inventions that have reshaped the world: the printing press, gunpowder, and the compass. Printing gives one power over one's people through propaganda, gunpowder over other peoples through force, and the compass over the world through navigation. What role do science and mathematics play in the expansion of humanity's dominion over the cosmos? Are there any risks to this use of science and mathematics?

4. Francis Bacon reflects, "For even as in the business of life, a man's disposition and secret workings of his mind and affections are better discovered when he is in trouble rather than at other times, so likewise the secrets of nature reveal themselves more readily under the vexations of art than when they do their own way." To what extent is the modern experimental method a vexing or torturing of nature? Are there moral limits to how much one may vex or torture nature? Is looking at nature

through mathematical models a kind of vexing or torturing of nature?

5. René Descartes calls for the foundation of a new practical philosophy, "by means of which, knowing the force and actions of fire, water, air, the stars, the heavens, and all other bodies that surround us, just as distinctly we know the various skills of our craftsmen, we might be able, in the same way, to use them for all the purposes for which they are appropriate, and thus render ourselves, as it were, masters and possessors of nature." Is scientific mastery and possession of nature an appropriate ambition for humanity? Are there limits to our mastery and possession?

6. In Michael Crichton's *Jurassic Park*, Dr. Ian Malcolm reflects, "Scientists are actually preoccupied with accomplishment. So they are focused on whether they can do something. They never stop to ask if they should do something." Keyser celebrates scientists and mathematicians of accomplishment. Is there room in his pantheon for the concerns of Dr. Malcolm?

7. Much of our thought and speech about the universe is framed by mathematical models. In what manner are these models icons, i.e., images through which truths about the universe are revealed? To what extent are these models idols, i.e., images that obstruct perception of the truth about the universe?

8. Keyser finds in Epicurus important roots for modern science, including "dogmas" such as "the denial of divine interference." To what extent is science dogmatic? Are dogmatism and being dogmatic the same? Is dogma good for science? Is there a tension between dogma and free thought?

9. To what extent are research and science political? What responsibility do political leaders have to further science? What authority do scientific leaders have in the body politic?

10. Keyser recognizes wonder as "the great question-asker." What is the difference between wondering about something, e.g., when one is struck and moved to ask questions, and contemplative wonder, e.g., when one gazes in wonder at fireworks or at a sunset. Are these different forms of wonder? Is one more fundamental than or superior to the other? Is one more central to science?

REFERENCES

For the quotation preceding the Preface, see C.J.K., *The Pastures of Wonder: The Realm of Mathematics and the Realm of Science* (Columbia University Press, 1929), p. ix. The text of the twelve primary chapters is extracted from C.J.K., *Portraits of Famous Philosophers Who Were Also Mathematicians* (Scripta Mathematica, 1939).

All references to "Diogenes" refer to Diogenes Laërtius, *Lives of Eminent Philosophers*, available at Perseus Digital Library, Tufts University.

INTRODUCTION

The first section of the Introduction is excerpted from C.J.K., *Mathematical Philosophy: A Study of Fate and Freedom* (E. P. Dutton & Co., 1922), pp. 135-137, 144-146, 150-152. The final three paragraphs are extracted from the Foreword to C.J.K., *Portraits*.

C.J.K. refers the reader to C.J..K, "The Role of Infinity in the Cosmology of Epicurus," *Scripta Mathematica*, Vol. IV, No. 3 (July 1936).

"older than the sun or the sky": Gilbert Murray, *Four Stages of the Greek Religion* (Columbia University Press, 1912), p. 98n.

"Mathematics and philosophy are incompatible": Similar: "I once heard a distinguished mathematician say that the study of mathematics acts on the metaphysical instinct like sulphur on the itch." C.J.K., "Some Outstanding Problems for Philosophy," *The Journal of Philosophy: Psychology and Scientific Methods*, Vol. II, No. 8 (April 13, 1905), pp. 207-213; p. 207.

like Lucian of a much later time: Lucian of Samosata, *Hermotimus, or the Rival Philosophies* (or Philosophical Sects), 74, translated by H. W. Fowler and F. G. Fowler, Vol. II (Clarendon Press, 1905), p. 84.

PORTRAIT 1: PYTHAGORAS

"a mass of incredible fiction": John Burnet, *Early Greek Philosophy* (1920), p. 87. "When we come to the *Lives of Pythagoras*...we find ourselves once more in the region of the miraculous. They are based on authorities of a very suspicious character, and the result is a mass of incredible fiction."

"A [certain] way of life": Plato, *Republic*, Book X, 600b. "Pythagoras was himself especially honored for this, and his successors, even to this day, denominating a certain way of life the Pythagorean."

By Aristotle he is mentioned twice: *The Complete Works of Aristotle*, edited by Jonathan Barnes, Vol. II (Princeton University Press, 1984), pp. 2441-2442.

He impiously laid claim: Iamblichus, *Life of Pythagoras*, translated by Thomas Taylor (1818), p. 65. "[I]n consequence of having divulged and described the method of forming a sphere from twelve pentagons, he perished in the sea, as an impious person, but obtained the renown of having made the discovery."

By Oswald Spengler the character: Oswald Spengler, *The Decline of the West: Perspectives of World History*, Vol. 2 (George Allen & Unwin, 1922), pp. 303-304. "Pythagoras, Mohammed, and Cromwell embody one and the same movement in three Cultures."

A speculation found later in the Timaeus: Plato, *Timaeus*, translated by W. R. M. Lamb (Harvard University Press, 1925), 37d. "[A]s He set in order the Heaven, of that Eternity which abides in unity He made an eternal image, moving according to number, even that which we have named Time."

PORTRAIT 2: PLATO

C.J.K. cites Thomas Little Heath, *A History of Greek Mathematics*, Vol. 1 (Clarendon Press, 1921), pp. 141-169; Andrew Seth Pringle-Pattison and George Johnston Allman, "Pythagoras," *Encyclopedia Britannica* (1911); Allman, *Greek Geometry, from Thales to Euclid* (Dublin University Press, 1877); and John Burnet, *Early Greek Philosophy* (1920), pp. 276-309.

"denying the gods": Diogenes, Book II.

"the wisest, the justest, and best of men": Plato, *Phaedo*, last sentence.

Plato used to say: *Plutarch's Lives*, Vol. III, edited by A. H. Clough (Little, Brown, & Co., 1861), p. 102. "Plato, when he saw his death approaching, thanked the guiding providence and fortune of his life, first, that he was born a man and a Grecian, not a barbarian or a brute, and next, that he happened to live in Socrates's age."

[Gibbon quotation in footnote]: *The History of the Decline and Fall of the Roman Empire*, edited by J. B. Bury (Methuen & Co., 1908), p. 266.

he teaches you nothing: Joseph Joubert, *Joubert: A Selection from his Thoughts*, translated by Katharine Lyttleton (Duckworth and Co., 1898), Chapter XXIII, p. 181. "Plato shows us nothing, but he brings brightness with him; he puts light into our eyes, and fills us with a clearness by which all objects afterwards become illuminated. He teaches us nothing; but he prepares us, fashions us, and makes us ready to know all."

Son of Apollo: Diogenes, Book III.

The good, the true, and the beautiful: The triad of transcendentals. Victor Cousin, *Lectures of the True, the Beautiful, and the Good*, translated by O. W. Wight (D. Appleton and Company, 1854).

"Plato is philosophy, and philosophy Plato": *The Works of Ralph Waldo Emerson*, Vol. IV (Macmillan and Company, 1883) p. 284. "Plato is philosophy, and philosophy Plato, at once the glory and the shame of mankind, since neither Saxon nor Roman has availed to add any idea to his categories."

[Whitehead quotation in footnote]: *Process and Reality: An Essay in Cosmology* (Macmillan and Co., 1929,) p. 63.

PORTRAIT 3: ARISTOTLE

C.J.K. lists twenty-one of Aristotle's works, grouping them into four categories: Logical, Scientific, Esthetic, and Philosophical.

"Master of those who know": Dante, *Inferno*, Canto IV, 131. "Maestro di color che sanno."

"the house of the reader": *De Vita Aristotelis*, a Greek manuscript with Latin translation, disputed authorship.

for though Euripides was the first: Gilbert Murray, *A History of Ancient Greek Literature* (William Heinemann, 1897), p. 252. "The poet lived, so Philochorus says, on his own estate at Salamis, and worked in a cave facing the sea.…[H]e was the first Greek who collected a library, the writer and thinker, not the man of affairs."

His motto was nil admirar: Aristotle, *Nicomachean Ethics*, Book IV, Chapter 3. "Nothing is great to him."

Aristotle's will: Diogenes, Book V, Chapter 1.

PORTRAIT 4: EPICURUS

C.J.K. cites Diogenes Laërtius, *Lives of Eminent Philosophers*, translated by Robert Drew Hicks (Harvard University Press, 1925); John Mason, *Lucretius: Epicurean and Poet* (John Murray, 1907); and Cyril Bailey, *The Greek Atomists and Epicurus* (Clarendon Press, 1928); and three translations of Lucretius, *De Rerum Natura*, translated by H. A. J. Munro (Deighton Bell and Co., 1864), Cyril Bailey (Clarendon Press, 1910), and William Henry Denham Rouse (Harvard University Press, 1924).

C.J.K. again refers the reader to C.J.K., "The Role of Infinity in the Cosmology of Epicurus," *Scripta Mathematica*, Vol. IV, No. 3 (July 1936).

"were enough to fill many cities": William Wallace, "Epicurus," *Encyclopedia Britannica*, Ninth Edition, Vol. VIII (Samuel L. Hall, 1878), pp. 472-476; p. 473.

[C.J.K. quotation in footnote]: Emphasis added. "Concerning the Figure and the Dimensions of the Universe of Space," *Science* (June 13, 1913), N.S. Vol. XXXVII. No. 966, pp. 885-895; p. 892. The quotation-within-quotation is inaccurate. Robert Green Ingersoll, "Oration on the Gods," *The Gods, and Other Lectures* (D. M. Bennett, 1876), pp. 1-60; p. 56. "The religion of our day, and country, is no more exempt from the sneer of the future than the others have been."

"When bodies are borne": *De Rerum Natura*, Book II, lines 217-220 (H. A. J. Munro translation, 1866) and lines 251-260 (W. H. D. Rouse translation, 1924).

[Sextus Empiricus quotation in footnote]: *Against the Philosophers*, Book I, lines 1-5, in *Sextus Empiricus*, Vol. IV, edited by R. G. Bury (Harvard University Press, 1944).

"It is impossible to live": The fifth in a series of "sovran [sovereign] maxims." Diogenes, Book X.

Marcus Aurelius endowed chairs: Eduard Zeller, *A History of Eclecticism in Greek Philosophy*, translated by S. F. Alleyne (Longmans, Green, and Co., 1883), pp. 189-193.

PORTRAIT 5: ROGER BACON

C.J.K. references *Opus Majus of Roger Bacon*, translated by Robert Belle Burke (University of Pennsylvania Press, 1928); Henry Osborn Taylor, *The Mediæval Mind* (Macmillan and Co., 1911); *Roger Bacon Essays*, edited by Andrew George Little (Clarendon Press, 1914); and William Romaine Newbold, *The Cipher of Roger Bacon*, edited by Roland Grubb Kent (University of Pennsylvania Press, 1928).

"suspected novelties": *Analecta Franciscana*, Vol. III (Quaracchi, 1897), p. 360, line 10. "Novitates suspectas."

"the greatest Apparition": Alexander von Humboldt, *Cosmos: Essai d'Une Description Physique du Monde*, Vol. II (Gide et J. Baudry, 1849), p. 300. "L'apparition la plus importante du moyen age."

"I have labored from my youth": Quoted in John K. Robertson, "The Birth of Modern Science," *The Scientific Monthly* (August 1926), pp. 138-151; pp. 143-144. For the original Latin, see Roger Bacon, *Opus Tertium*, Chapter XVII, in *Opera Quædam Hactenus Inedita*, Vol. I, edited by J. S. Brewer (Longman, Green, Longman, and Roberts, 1859), pp. 58-59.

"for ten years an exile": *Opus Tertium* (Brewer), Chapter I, p. 7. "Me jam a decem annis exulantem."

"to our beloved son": English reprinted in Taylor, p. 589. The original Latin is reprinted in *Opus Tertium* (Brewer), p. 1, and as Epistola CCCXVII in Edmundi Martene and Ursini Durand, *Thesaurus Novus Anecdtorum*, Vol. II (1717), p. 358.

"attacked dead and living worthies": Henry Osborn Taylor, *The Mediæval Mind*, Vol. II (Macmillan Company, 1919), p. 516.

"There are two modes": Most of this quotation is borrowed from *Opus Majus*, Vol. II, translated by Robert Belle Burke, pp. 583-584. The references to fire and an equilateral triangle are borrowed from Taylor, p. 506, which paraphrases Burke, Vol. II, p. 583: "For is a man who has never seen fire should prove by adequate reasoning that fire burns and injures things and destroys them, his mind would not be satisfied thereby, nor would he avoid fire, until he placed his hand or some combustible substance in the fire, so that he might prove by experience that which reasoning taught. But when he has had actual experience of combustion his mind is made certain and rests in the full light of truth. Therefore reasoning does not suffice, but experience does. This is also evident in mathematics, where proof is most convincing. But the mind of one who has the most convincing proof in regard to the equilateral triangle will never cleave to the conclusion without experience, nor will he heed it, but will disregard it until experience is offered him by the intersection of two circles, from either intersection of which two lines may be dawn to the extremities of the given line; but then the man accepts the conclusion without any question."

"The study of wisdom": It is curious that Keyser borrows this quote from Taylor, pp. 492-493, rather than Burke, p. 66: "[T]he study of wisdom can always continue in this life to increase, because nothing is perfect in human discoveries. Therefore we of a later age should supply what the ancients lacked, because we have entered into their labors, by which, unless we are dolts, we can be aroused to better things, since it is most wretched to be always using old discoveries and never be on the track of new ones..." The Latin is available in *Opus Majus*, edited by John Henry Bridges (Williams and Norgate, 1900), Supplemental Volume, Part II, Chapter XV, pp. 69-70.

PORTRAIT 6: DESCARTES

All references to *Discourse* refer to *The Discourse on Method*, printed in *French and English Philosophers*, Vol. 34 of the Harvard Classics, edited by Charles W. Eliot (P. F. Collier

& Son, 1910). All references to *Oeuvres* refer to *Oeuvres de Descartes*, edited by Charles Adams and Paul Tannery (Léopold Cerf).

C.J.K. lists five of Descartes' works: *Discourse on Method, Meditations, Principles of Philosophy, Treatise on the Passions of the Mind, and World.*

"Little philosopher": Adrien Baillet, *La Vie de M. Descartes* (1691), p. 16. "Qu'il coûtume d'appeller *son Philosophe*, a cause de la curiosité insatiable avec laquelle il luy demandoit les causes & les effets de tout ce qui luy passoit par les sens." [emphasis in original]

"ardently desirous of instruction": Descartes, *Discourse*, Part I, p. 7. "J'avais un extrême désir de les apprendre."

"tranquility and respose": Descartes, Letter DLXXXIV (15 Janvier 1650), *Oeuvres*, Vol. V (1903), p. 467, lines 20-23.

"The long chains": *Discourse*, Part II, p. 18.

"But I have just been at Leyden": Letter XLIX (Novembre 1633), *Oeuvres*, Vol. I (1897), p. 270, lines 10-17.

Sir Michael Foster's assertion: Foster, *Lectures on the History of Physiology* (Cambridge University Press, 1924), Lecture X, p. 266. "Descartes was a philosopher, not a physiologist. He took interest in the problems of the living body only so far as they bore on the greater problems of the why and the wherefore of the universe."

Thomas H. Huxley in his lecture: Huxley, "On the Hypothesis that Animals are Automata, and Its History," Collected Essays (D. Appleton and Company, 1901), pp. 199-250; pp. 201, 203.

"elegant trifling": William Wallace, "Descartes," *Encyclopedia Britannica*, Eleventh Edition, Vol. VIII (1910), pp. 79-90; p. 81. "Thus Descartes is a type of that spirit of science to which erudition and all the heritage of the past seem but elegant trifling."

"I am becoming daily more": *Discourse on Method*, Part VI, p. 60.

PORTRAIT 7: PASCAL

"People": Niccolò Machiavelli, *Il Principe* (The Prince), Chapter XXII (1532): "E perché sono di tre generazioni cervelli; l' uno intende per sé; l' altro discerne quello che altri intende; e il terzo non intende per se stesso, né per dimostrazione di altri."

Proposition 32: Book I in *The Thirteen Books of Euclid Elements*, translated by Thomas L. Heath (Cambridge University Press, 1908), p. 316.

"Helen of Geometry": Jean-Étienne Montucla, *Histoire des Mathématiques* (1799), Vol. II, Part IV, Liv. I, p, 55. "[L]e commencement des querelles nombreuses que cette Hélène des géomètres causa parmi eux."

The Provincials, "a triumph": George Saintsbury and George Chrystal, "Pascal," *Encyclopedia Britannica*, Ninth Edition, Vol. XVIII (1885), pp. 333-338; p. 336.

PORTRAIT 8: SPINOZA

"Woe to him": Delivered February 21, 1877, not in 1880, at the Hague. Ernest Renan, *Spinoza* (Michel Lévy Frères, 1877), p. 26. "Malheur à qui en passant enverrait l'injure à cette figure douce et pensive! Il serait puni comme tous les cœurs vulgaires par sa vulgarité même et par son impuissance à comprendre le divin. Lui, de son piédestal de granit, enseignera à tous la voie du bonheur qu'il a trouvé, et, dans des siècles, l'homme cultivé qui passera sur le Pavilioengragt dira en lui-même: 'C'est d'ici peut-être que Dieu a ete vu de plus près.'"

"People spoke of Spinoza": Gotthold Ephraim Lessing, German philosopher, in conversation with Carl Gustav Jacob Jacobi. For the German, see Jacobi, *Ueber die Lehre des Spinoza in Briefen an den Herrn Moses Mendelssohn* (1785), p. 27. "Reden die leute doch immer von Spinoza wie von einem todten hunde." For the English, see Jacobi, "Concerning the Doctrine of Spinoza in Letters to Herr Moses Mendelssohn," *The Main Philosophical Writings and the Novel* Allwill, translated by George di Giovanni (McGill-Queen's University Press, 1994), pp. 173-251; p. 93.

"the holy and excommunicated": Quoted in Andrew Seth Pringle-Pattison. For the German, see Friedrich Schleiermacher, "Über das Wesen der Religion," Über die Religion (Johann Friedrich Unger, 1799), pp. 38-133; p. 54. For the English, see Schleiermacher, "Second Speech: The Nature of Religion," *On Religion: Speeches to Its Cultured Despisers*, translated by John Oman (Harper & Row, 1958), pp. 26-118; p. 40.

"Something far more deeply": William Wordsworth, "Lines, composed a few miles above Tintern Abbey, on revisiting the banks of the Wye during a tour—July 13, 1798," in *Lyrical Ballads, with Other Poems*, Vol. I (Longman and O. Rees, 1800), pp. 201-210. Most of the poems were written by Wordsworth, but co-author Samuel Taylor Coleridge contributed several, including "The Rime of the Ancyent Marinere."

"no one would be molested": Article XIII, Union of Utrecht (January 23, 1579).

For sheer malignity the curses: An English version is reprinted in *The Oldest Biography of Spinoza*, edited by Abraham Wolf (Dial Press, 1927), pp. 145-147.

"an adamantine conviction": Spinoza, *Treatise on the Correction of the Understanding (Tractatus de Intellectus Emendatione)*, paragraph 13. "For the greatest good is for him to attain to the enjoyment of such a nature together with other individuals, if this can be. What is that nature I shall show in its proper place, namely, that it is the knowledge of the union which the mind has with the whole of nature."

The tender, which was in: Letter from J. Ludovicus Fabritius to Spinoza, reprinted as Letter LIII (February 16, 1673) in R. Willis, *Benedict de Spinoza: is Life, Corre-*

spondence, and Ethics (Trübner & Co., 1870), pp. 365-366. "You would enjoy the most perfect freedom in philosophizing, which His Highness feels assured you would not abuse by calling in question the established religion of the state."

teaching in public would be incompatible: Letter from Spinoza to J. Louis Fabritius, reprinted as Letter LIV (March 30, 1673), pp. 366-367. in R. Willis, *Benedict de Spinoza: is Life, Correspondence, and Ethics* (Trübner & Co., 1870), pp. 365-366.

"eloquently reasoned defense": Andrew Seth Pringle-Pattison, "Spinoza," *Encyclopedia Britannica*, Eleventh Edition, Vol. XXV (1911), pp. 687-691; p. 690.

"the Intellectual love of God": Spinoza, *Ethics*, Part V, Corollary to Proposition XXXII, (G. P. Putnam's Sons, 1888), pp. 327-328; p. 328. "From the third kind of knowledge necessarily arises the Intellectual Love of God. For, from this kind of knowledge arises perfect joy associated with the idea of God as its cause; that is (by 6 Def. Pass.), Love of God, not as God is imagined to be present (by Prop. 29), but as God is understood to be eternal; and this is what I call the intellectual love of God."

PORTRAIT 9: LEIBNIZ

C.J.K. refers to Herbert Wildon Carr, *Leibniz* (Ernest Benn Limited, 1929) and *The Monadology of Leibniz*, also edited by Carr (University of Southern California, 1930).

"very different things in view": William Ritchie Sorley, "Leibnitz," *Encyclopedia Britannica*, Eleventh Edition, Vol. XVI (1911), pp. 385-390; p. 385. The underlying source appears to be a memoir or "confessions" authored by Leibniz and reprinted in G. E. Guhrauer, *Gottfried Wilhelm Freiherr von Leibniz: Eine Biographie*, Vol. I (1846), p. 44: "in einer ganz anderen Richtung."

"the last of the universals": While it is possible De Quincey said this at some point, the phrase does not appear in his extant writings. However, he comes close in *Letters of*

De Quincey, the English Opium-Eater, to a Young Man whose Education has been Neglected (John Penington, 1843), pp. 26-27. "This Englishman [probably Samuel Taylor Coleridge] set out in life, as I conjecture, with a plan of study modelled upon that of Leibnitz—that is to say, he designed to make himself (as Leibnitz most truly was) a *Polyhistor* or Catholic student...Leibnitz, however anxious to throw out his mind upon the whole encyclopædia of human research, yet did not forget to pay the price at which only any *right* to be thus discursive can be earned: he sacrificed to the austerer muses: knowing that God geometrizes eternally, he rightly supposed that in the universal temple Mathesis must furnish the master key which would open most shrines."

[Gladstone quotations in footnote]: William Ewart Gladstone, On Books and the Housing of Them (Dodd, Mead & Company, 1890), p. 9. "Let us bow our heads..." *Gleanings of Past Years, 1843-78*, Vol. III (John Murray, 1879), p. 167. "As we now stand..."

"the most powerful instrument": William Benjamin Smith, *Infinitesimal Analysis*, Vol. I (Macmillan Company, 1898), p. Vii.

PORTRAIT 10: BERKELEY

All references to *Works* refer to *The Works of George Berkeley*, edited by Alexander Campbell Fraser (Clarendon Press, 1901). All references to *Misc.* refer to *Mathematical Miscellanies*, in *The Works of George Berkeley*, Vol. I, edited by George Sampson (George Bell and Sons, 1908).

C.J.K. lists eleven of Berkeley's works.

"Here is a fine young gentleman": Quoted in George-Monck Berkerley, *Poems*, edited by Eliza Berkeley (1797), p. ccclxxxi.

"I was distrustful": *Works*, Vol. I, pp. 1-92; p. 79.

"I am young, I am an upstart": Ibid., p. 14.

"I have known him": Thomas Blackwell, *Memoirs of the Court of Augustus*, Vol. II (Edinburgh, 1755), pp. 277-278.

"passed with great distinction": Leslie Stephen, "Berkeley," *Dictionary of National Biography*, Vol. IV (Macmillan and Co., 1885), pp. 348-356; p. 348.

"the highest pinnacle of human knowledge": "On the Algebraic Game," *Misc.*, pp. 51-58; p. 57. *Works*, Vol. III, pp. 54-62; p. 60.

"the clear light of demonstration": "Arithmetic Demonstrated Without Euclid or Algebra," *Misc.*, pp. 1-38; p. 5. Latin translation provided by *Works*, Vol. III, pp. 3-38; p. 6. "In clara tamen demonstrationum."

"abstract[ing] the mind": "On the Algebraic Game," *Misc.*, p. 57. "[S]o mathematical studies, besides their proper aims and uses, have that collateral one, that they abstract the mind from the senses, sharpen and confirm the talents." *Works*, Vol. III, pp. 54-62; p. 61. "Sic disciplinæ mathematicæ, præter fines ac usus singulis proprios, illud etiam collaterale habent, quod mentum a sensibus abstrahant, ingeniumque acuant et figant."

"If with a view": Berkeley, *The Analyst*, in Works, Vol. III, pp. 17-60; p. 25.

"the land of the setting sun": The phrase was widely used to refer to the New World. Berkeley's "A Proposal" (1725), concerning the Bermuda college, is reprinted *Works*, Vol. III, pp. 341-362. See also "Verses," p. 235-236.

"I could wish that all the things": C.J.K. has reversed the chronological order of these two letters from Berkeley. The first quotation is from Works, Vol. II, pp. 19-21; p. 20.

"What you have seen": Ibid., pp. 15-18; p. 18.

"Some truths there are": *A Treatise Concerning the Principles of Human Knowledge*, in *Works*, Vol. I, pp. 211-347; pp. 260-261.

"When I deny sensible things": *Three Dialogues Between Hylas and Philonous*, in *Works*, Vol. I, pp. 349-485; p. 446-447.

PORTRAIT 11: KANT

All references to CP refer to *Collected Papers of Charles Sanders Peirce*, edited by Charles Hartshorne and Paul Weiss (Harvard University Press), and use the paragraph numbering system.

C.J.K. lists eight of Kant's works.

"a woman of exalted character": Thomas De Quincey, *Last Days of Immanuel Kant and Other Writings* (Adam and Charles Black, 1862), p. 103.

"[he] never spoke of his mother" – Ibid.

printed his first paper: *Thoughts on the True Estimation of Living Forces* (1747).

Herbert Spencer's candid: Herbert Spencer, "The Ethics of Kant," *Essays: Scientific, Political, & Speculative*, Vol. III (Williams and Norgate, 1891), pp. 192-216; p. 206.

"two hours a day": CP (1931) 1.4.

"the king of modern thought": CP (1931) 1.369.

"One of his practical principles": Will Durant, *The Story of Philosophy* (Garden City Publishing, 1927), p. 286. C.J.K. omits quotation marks around this sentence.

the following couplet: Lord Houghton, "On Certain Present Phenomena of the Imagination," *The Fortnightly Review*, Vol. XXV, No. 145 (January 1, 1879), pp. 62-79; p. 76. A note explains that this address was delivered to the Leeds Philosophical Society on December 17, 1878. Mysteriously, W. K. Clifford refers to Houghton's "perfect" couplet more than a year earlier in "Cosmic Emotion," *The Nineteenth Century: A Monthly Review*, Vol. II (October 1877), pp. 411-429; p. 411. It is possible Houghton presented his couplet privately prior to his public address.

"Act always on such a maxim": *Fundamental Principles of the Metaphysics of Morals*, translated by T. K. Abbott, in *Literary and Philosophical Essays*, Harvard Classics, edited by Charles W. Eliot (Collier & Son, 1910), pp. 305-317; p. 368. For the German, see *Grundlegung zur Metaphysik der Sitten* (1785), p. 42. "Der kategorische Imperativ ist also nur ein einziger und zwar dieser: handle nur nach derjenigen Maxime, durch die du zugleich wollen kannst, daß sie ein allgemeines Gesetz werde."

PORTRAIT 12: PEIRCE

All references to CP refer to *Collected Papers of Charles Sanders Peirce*, edited by Charles Hartshorne and Paul Weiss (Harvard University Press), and use the paragraph numbering system. C.J.K. references this series.

"He educated me": Quoted in Paul Weiss, "Peirce, Charles Sanders," *Dictionary of American Biography*, Vol. 14, edited by Dumas Malone (Charles Scribner's Sons, 1934), pp. 398-403; p. 398. Others cite Manuscript 1608 as this quotation's original source.

"the science which draws": Benjamin Peirce, *Linear Associative Algebra* (1870), p. 2.

"I am a man": CP (1931) 1.10.

"[A]ll my life": CP (1933) 4.118.

Joseph Jastrow: C.J.K. provides the full citation in-line, which has been moved here for readability. Joseph Jastrow, "The Widow of Charles S. Peirce," *Science*, Vol. 80, No. 2081 (November 16, 1934), pp. 440-441.

"Consider what effects": CP (1934) 5.402.

"In order to ascertain": CP (1934) 5.9.

"[T]o be deep": CP (1934) 5.17.

"So then, the writer": CP (1934) 5.414.

"syllogism of transposed quantity": CP (1934) 6.114-115.

"Every Hottentot": CP (1958) 7.552.

"as if a geographer": CP (1933) 3.415. "For this subject, although always recognised as an integral part of logic, has been left untouched on account of its intricacy. It is as though a geographer, finding the whole United States, its topography, its population, its industries, etc., too vast for convenient treatment, were to content himself with a description of Nantucket. This comparison hardly, if at all, exaggerates the inadequacy of a theory of reasoning that takes no account of relative terms."

"It is the conception of a beginningless": C.J.K. references CP 1.406-416, describing this "first sketch of the idea" as a "veritable sunburst of scientifico-speculative genius."

CONCLUSION

The conclusion is excerpted from C.J.K., "The Human Significance of Mathematics," *Science*, New Series, Vol. 42, No. 1089 (November 12, 1915), pp. 663-680; pp. 675-676.

"the blind hurry of the universe": Bertrand Russell, "The Free-Man's Worship," *The Independent Review*, Vol. I, No. 3 (December 1903), pp. 415-24; p. 422.

"The past rises before": Robert. G. Ingersoll, *The Ghosts and Other Lectures* (C. P. Farrell, 1892), pp. 229-232, excerpted from a speech—eulogy to the fallen of the U.S. Civil War—at the Soldiers' Reunion in Indianapolis on September 21, 1876. Ingersoll's original is "rises before me," whereas C.J.K. changes it to "rises before us."

bonitas, unitas, infinitas, immutabilitas: St. Thomas Aquinas, *Summa Theologica*, Vol. I: Question 6 for bonitas (goodness); Question 11 for unitas (oneness or unity); Question 7 for infinitas (infiniteness); and Question 9 for immutabilitas (immutability).

"the compulsion in the way": Gilbert Murray, *Four Stages of Greek Religion* (Columbia University Press, 1912), p. 115, from where much of this clause is paraphrased. Murray translates the term ἀνάγκη φύσεος as "the compulsion in the way things grow." He also references Euripides, *The Trojan Women*, line 886, as an example of its usage. *Euripidis Fabulae*, Tomus II, edited by Gilbert Murray (Clarendon Press, 1904), for the Greek, and *The Trojan Women of Euripides*, translated by Gilbert Murray (George Allen & Unwin, 1905), pp. 54-55, for the English.

a fine thread: Murray, *Four Stages of Greek Religion*, p. 115, attributes the "fine thread" image to Zeno of Citium (not to be confused with Zeno of Elea, who devised famous thought experiments concerning infinity).

[the apeiron] of Anaximander: Anaximander's writings are preserved only in fragments, such as this definition of the apeiron transmitted to us from Theophrastus by way of Simplicius: "from which arise all the heavens and the worlds within them." John Burnet, *Early Greek Philosophy* (Adam and Charles Black, 1892), p. 49.

cosmic order...of Heracleitus: W. Windelband, *A History of Philosophy*, translated by James H. Tufts (Macmillan Company, 1898), p. 36. "This rhythm of events...is therefore the only permanent; it is termed by Heraclitus the destiny, the order, the reason of the world."

space-filling sphere...of Parmenides: Burnet, pp. 185-86. "One path only is left for us to speak of, namely, that *It is*. In it are very many tokens that what is, is uncreated and indestructible, alone, complete, immovable and without end...Wherefore it is not permitted to what is to be infinite; for it is in need of nothing; while, if it were infinite, it would stand in need of everything."

four material and two psychic: Burnet, p. 222. "At one time things grew together to be one only out of many, at another they parted asunder so as to be many instead of one—Fire and Water and Earth and the mighty height of Air; dread Strife, too, apart from these and balancing every one of them, and Love among them, their equal in length and breadth."

infinitude...of Anaxagoras: Burnet, p. 283. "All other things partake in a portion of everything, while Nous is infinite and self-ruled, and is mixed with nothing, but is alone, by itself...For it is the thinnest of all things and the purest, and it has knowledge about everything and the greatest strength."

"seeds of things": Lucretius, *De Rerum Natura* (On the Nature of Things), Book I, lines 58-59, translated by H. A. J. Munro (George Bell and Sons, 1900), p. 2. "[T]o call matter and begetting bodies of things and to name seeds of things and also to term first bodies, because from them as first elements all things are."

Cosmic Year: Keyser, *Mathematical Philosophy*, p. 228. "[T]he philosophy of the cosmic cycle regards all the changes of which the universe is capable as constituting an immense but finite and closed system of transformations, which follow each other in definite succession, like the spokes of a gigantic revolving wheel, until all possible changes have occurred in the lapse of a long but finite period of time—called a cosmic cycle or a cosmic year—whereupon everything is repeated precisely, and so on and on without end."

Absolute of the Hegelian school: *The Science of Logic*, 21(b), in *The Logic of Hegel*, translated by William Wallace (Clarendon Press, 1892), p. 43. "Religion leads us to a universal, which embraces all else within itself, to an Absolute by which all else is brought into being: and this Absolute is an object not of the senses but of the mind and of thought."

THE SCIENCE VENERABLE

Voltaire once remarked—"One merit of poetry few will deny; it says more and in fewer words than prose." With equal significance we may say, "One merit of mathematics few will deny; it says more and in fewer words than any other science." The formula $e^{\pi i} = -1$ expresses a world of thought, of truth, of poetry and of religious spirit, for "God eternally geometrizes."

—Creed of David Eugene Smith

Made in the USA
Monee, IL
04 November 2026

67375846R00088